# Television Commercial Processes and Procedures

# Multiple Camera Video Series
By Robert J. Schihl

Television Commercial Processes and Procedures

Talk Show and Entertainment Program Processes and Procedures

Studio Drama Processes and Procedures

TV Newscast Processes and Procedures

# Television Commercial Processes and Procedures

Robert J. Schihl, Ph.D.

**Focal Press**

Boston London

**Library of Congress Cataloging-in-Publication Data**

Schihl, Robert J.
  Television commercial processes and procedures / Robert
  J. Schihl.
      p.  cm.—(Multiple camera video series)
  Includes bibliographical references (p.    ) and index.
  ISBN 0-240-80098-2 (pbk.)
  1. Television—Production and direction.  2. Television
advertising.  I. Title.  II. Series: Schihl, Robert J.
Multiple camera video series.
PN1992.75.S36  1991
791.45'0233—dc20                              91-16864
                                              CIP

**British Library Cataloguing in Publication Data**

Schihl, Robert J.
  Television commercial processes and procedures.
  (Multiple camera video series)
  I. Title  II. Series
  659.143

  ISBN 0-240-80098-2

Butterworth–Heinemann
80 Montvale Avenue
Stoneham, MA 02180

10  9  8  7  6  5  4  3  2  1

Printed in the United States of America

# Contents

# Preface

*Multiple camera video production* refers to those varying video projects that create the products of the multiple camera video technology. It is usually the video production associated with a studio setting. Some of the most common multiple camera studio production today includes television commercial production, newscast production, drama production (mainly soap operas), and talk show/entertainment production. Of these, only the daily newscast and occasional television commercial are still being produced at local television production studios. Occasionally, public service talk shows are produced at the local facility.

In network television studio production, besides the daily newscasts, the continuing daily soap opera production still occupies one of the busiest and most productive of the studio projects.

While the basic skills necessary for studio production remain the same across studio production genres, pre-production tasks and production roles vary immensely from one genre to another. The beginning television producer, director, and production crews need to be aware of those studio production elements that are similar across genres as well as those that differ. It is not uncommon to find well-meaning beginners performing poorly in their roles across genres when principles and criteria for one genre were erroneously applied to another. Professionally, even apparently similar studio production roles across television program genres require a different mind set and, consequently, different skills application. Hence, at beginning and intermediate television production levels, awareness of the various genres and their differences encourages varying skill development and consequent employment marketability.

Academic goals for beginning television production skills are usually oriented to some form of studio operation. Therein students have the opportunity to learn all aspects of video production from the technology of the medium to the aesthetics of the final product. In some academic programs, knowledge of these concepts and skills is required before advancing to the single camera video experience. The television studio experience encompasses all phases of video *producing skills* from those of the producer to those of the director and all phases of video *production technology* from that of the camera operator (video camera skills) to that of the technical director (video editing skills). Given success in the learning stages of video production, skills still need expression. The most common expression of video production skills is to engage in some form of television studio production. Beyond a basic television studio production handbook, there is not an adequate text available to design and organize differing studio production genres. This book concentrates on producing and the production of the television studio—three camera—commercial.

The growth of both broadcast and cable television in the United States and abroad in the 1990s shows ever increasing development and expansion. The areas of low power television (LPTV) production, the continuing increase in cable origination programming, and the introduction of high definition television (HDTV) attest to that development and expansion. The rapid fragmentation of cable and satellite television in general is spurring the proliferation of independent television production facilities and the multiplication of television studio program genres. Producers for these independently produced and syndicated programs seek from media conventions the guidelines for producing and the production of the various genres. Publishers of books on the subject of television are seeing a growing need in these areas and are responding with an increasing number of publications. These publications assist the entrepreneur in the development of video production houses and in the training of producers. What is needed by both the broadcast educator as well as the video production entrepreneur is guidance for producing and production procedures, information on video production organizational flow, and the definition of video producing and production roles.

Following the publication of my earlier book *Single Camera Video: From Concept to Edited Master* (Focal Press, 1989), I found that educators, beginning professionals, and industry reviewers alike applauded the format in which I presented the procedural, organizational, and role definition stages of the single camera video technology. I was asked why such a format was not available for multiple camera video production. Having spent more years both in broadcast education and professional production with multiple camera video production than with single camera video production, I asked myself the same question. I hope this text is the desired response.

*Television Commercial Processes and Procedures* is intended to meet the needs of those individuals from intermediate television education to the beginning television production facility entrepreneur who needs a clear and comprehensive road map to organizing a television studio commercial production. This book presents the conceptual preproduction stages of commercial production (e.g., working with a client; writing a script; auditioning a cast; and constructing a production budget, studio set design, and lighting and audio design) as well as all levels of production skill definition and organization until the director calls for a wrap to the studio production. This road map includes the thoroughness of a flowchart for all preproduction, production, and postproduction personnel roles and the specific detail of preproduction, production, and postproduction organizing forms that facilitate and expedite most roles throughout the process of the studio production of the television commercial.

In *Single Camera Video: From Concept to Edited Master* I referred to the approach I took as "a cookbook recipe approach." I have not only not regretted the comparison but have grown confirmed in the analogy. Similar to single camera video production, multiple camera video production is a process needing the guidance of a good kitchen recipe—an ordered, procedural, and detailed set of guidelines. As with a good recipe, after initial success, a good cook adjusts ingredients, eliminates some, adds others, and makes substitutions according to taste and experience with the recipe. So, too, is it with the approach taken in this text. The production organizing details covered in this text are such that after initial introduction, depending often on available personnel, studio equipment, or production time frame, preproduction and production task roles can be multiplied, combined, or eliminated—just like the ingredients in a favorite recipe.

Specifically, the approach that this text takes to multiple camera video production of the television commercial is one of organization. The most frequent and vociferous compliment to *Single Camera Video* is that it organizes the whole gamut of the preproduction, production, and postproduction of single camera video production. In the same way, this text is designed to organize the multiple camera video production of the television studio commercial. Books on the other television studio genres (the television newscast, the talk show/entertainment program, and the television studio drama production) are also available from Focal Press in worktexts similar to this.

The process of television commercial production is presented in Chapter 1 in flowchart checklist form for preproduction, production, and postproduction stages of production. For each of the three stages, producing and production personnel roles are presented, with responsibilities and obligations listed in their order of accomplishment. In Chapter 2, the processes of preproduction, production, and postproduction are presented in chronological order of production role performance. In Chapter 3, production organization forms are provided for various preproduction and production tasks. These forms are designed to facilitate the goal of producing and production requirements at certain stages of the process of producing and production. A description of and glossary for each form are presented in Chapter 4 to facilitate use of the forms.

In determining the method of presenting the television studio commercial production genre in this text, some choices had to be made. Production procedures had to be adapted to a learning process. The first choice was to decide the number of producing personnel and the studio production crew size. The number of producing roles and studio production crew members usually is determined by the production budget and the number of available personnel (i.e., students) in an academic environment. The choice was made to create above-the-line and below-the-line staff and crew for an average size production with a minimum number of personnel. As in the analogy of a good recipe, combine, subtract, or multiply staff and crew members according to costs and availability. The industry does that, too.

Another choice was to determine the level of studio hardware and technology sophistication to assume in television commercial production. No two television studio facilities are alike, especially in academic environments. The criterion for level of hardware and technology sophistication assumed was to be aware that no matter how much or how little a studio facility has in terms of hardware and technology, the students leaving their training time in any type of production facility still have to know some point of minimal reference to the "real world" of television commercial production and employment skill expectation standards. A teacher of broadcasting can easily structure this text and its approach to an individual facility. Without an ideal point of reference, it would be very difficult to structure up to a level that was not covered in this text. Students must be made aware of what to expect in a "normal" or "ideal" production facility and from a studio production crew. This is what I have tried to convey. I have taught studio commercial production from a minimal facility to a full blown state-of-the-art facility. This text is adaptable to either.

I repeat what I wrote in *Single Camera Video*. The video production industry or television studio commercial production is not standardized in the procedural manner in which its video products are created. This text, although it can appear to recommend standardized approaches to achieving video products, is not intended to imply that the industry or television commercial production is standardized nor is it an attempt to standardize the industry or commercial production. The presentation of ordered production steps is merely academic; it is a way of teaching and learning required stages in video production. The video product is the result of a creative process and should remain that way. This text provides order and organization to television studio commercial production for beginning producers and directors and for intermediate television teachers and students of video production. Once someone becomes familiar with the process of any television commercial production, that person is encouraged to use what works and facilitates a task and to drop or rethink those elements that do not work or no longer facilitate the task.

I am currently a full professor—both a charter and senior television faculty member—at Regent University on the grounds of the Christian Broadcasting Network, home of the Family Channel. I teach in both the School of Radio, Television, and Film, and the School of Journalism, and work with faculty and graduate students of the Institute of Performing Arts.

## ACKNOWLEDGMENTS

I am indebted to many people over many years who have contributed unknowingly to this book:

To Joanne, Joel, and Jonathan for putting up with the interminable clacking of computer keys and the whine of a printer;

To my parents, Harold and Lucille, who unwittingly turned me on to television production in 1950 by purchasing our first television set;

To Dr. Marilyn Stahlka Watt, Chair, Department of Communication, Canisius College, Buffalo, New York, for introducing me to television broadcasting and for making me a television producer;

To Edward Herbert and Kurt Eichsteadt, Taft Broadcasting, for giving me the chance to have my own live, local prime-time television program;

To Marion P. Robertson, Chief Executive Officer, the Family Channel, Virginia Beach, Virginia, for giving me the opportunity to work and teach in a state-of-the-art national network television facility;

To Dean David Clark, Provost George Selig, and President Bob Slosser, Regent University, Virginia Beach, Virginia, for granting me the sabbatical to write this book;

To Karen Speerstra, Senior Editor, and Philip Sutherland, Acquisitions Editor, Focal Press, Stoneham, Massachusetts, for being the most encouraging editors and friends an author could have;

To Rob Cody, Project Manager; Julie Blim, "700 Club" Producer; and John Loiseides, Photographer, Christian Broadcasting Network, Virginia Beach, Virginia, for their research assistance;

To my thousands of television students from the State University of New York at Buffalo, the State University of New York College at Buffalo, Hampton University, and Regent University, and especially to those among them whose names I see regularly on the closing credits of network and affiliate television programs—for the thrill of seeing their names.

*A.M.D.G.*
*Robert J. Schihl, Ph.D.*
*Virginia Beach, Virginia*
*1991*

# Key to the Book

Creating a successful television studio production requires that many tasks at varying stages of production be performed in sequential order. Most tasks build upon one another and are interrelated with the tasks of other production personnel. The television industry has clearly defined roles for each production personnel member. This book provides a blueprint of the duties assigned to each production role. The duties of each role—producer, director, camera operator, etc.—are first presented in a checklist flowchart. Role duties are then cross-referenced within the text on the basis of studio production stages. The tasks necessary for every production role are divided and arranged at each stage of the production process. The flowchart is divided into the three chronological stages of television studio production: preproduction, production, and postproduction.

This book can be used as a text or reference. The reader can gain a comprehensive understanding of the production process by reading the entire book. The reader also can check specific personnel responsibilities or use particular forms.

Following is a listing of the personnel abbreviations used throughout this text.

| Personnel | Abbreviation |
|---|---|
| Producer | P |
| Director | D |
| Camera operators | CO |
| Audio director | A |
| Lighting director | LD |
| Continuity person | CP |
| Production assistant | PA |
| Microphone boom grip(s)/operator(s) | MBG/O |
| Video engineer | VEG |
| Assistant director | AD |
| Floor director | F |
| Teleprompter operator | TO |
| Technical director | TD |
| Videotape recorder operator | VTRO |
| Talent/actor/actress | T |
| Set designer | SD |
| Properties master/mistress | PM |
| Costume master/mistress | CM |
| Make-up artist | MA |

Chapter 1 presents a flowchart for preproduction, production, and postproduction stages by personnel role and in sequential task order (see figure). Chapter 2 details each production task in the order in which it is to be completed. Chapter 3 provides production organizing forms. Chapter 4 is a key to terms and information for the production forms in Chapter 3.

---

Choose a television studio production stage. Here, the PREPRODUCTION stage was chosen.

Each television studio production role is abbreviated (e.g., P) and sequential production tasks numbered (e.g., (1), (2), (3), etc.). To locate the fifth PREPRODUCTION task responsibility of a PRODUCER (P) look down the checklist for (5).

Every television studio production role and task is listed here and is explained in detail in Chapter 2. The page number following the entry directs you to the explanation. Chapter 2 will also aid you in learning the responsibilities of the other crew members.

Reading this information will indicate if there is a production organizing form available in Chapter 3 to assist you in a particular task.

Each television studio production organizing form is explained in detail in a glossary of terms and information required for the form in Chapter 4.

**PREPRODUCTION**

**PRODUCER (P)**

- ☐ (1) Contacts/is contacted by the client/agency  11
- ☐ (2) Researches/reviews marketing problem (marketing analysis form)  11
- ☐ (3) Writes a commercial script (video script form)  11
- ☐ (4) Designs the storyboard (storyboard form)  11
- ☐ (5) Constructs a budget (production budget form)  11
- ☐ (6) Arranges for the production facility/studio  12
- ☐ (7) Meets in a preproduction meeting with the client/agency  12
- ☐ (8) Chooses/meets with the director  13
- ☐ (9) Creates the production schedule (production schedule form)  14
- ☐ (10) Holds auditions; supervises casting the commercial with the director  14
- ☐ (11) Approves set(s) designs; authorizes expenses and purchase requisitions  16
- ☐ (12) Approves properties list; authorizes expenses and purchase requisitions  17
- ☐ (13) Approves costume designs; authorizes expenses and purchase requisitions  17
- ☐ (14) Approves make-up design; authorizes expenses and purchase requisitions  18
- ☐ (15) Approves audio plot design and music; authorizes expenses and purchase requisitions  23

**Annotated flowchart.**

# Television Commercial Production

## INTRODUCTION

At many network affiliate television stations, the newscast may be the facility's only live and local origination video production. Often the only other use of the facility's production studio is for the creation of local commercials and public service announcements. There are two major types of commercials seen on television today: the high quality, big budget, national advertising agency produced commercial for a major national client and the locally produced, in-house commercial.

The nationally produced commercial for big name products is usually a single camera film production. The film is often transferred to video and edited. The national advertising agency commercials are known for their creativity and innovation as well as for their quarter million dollar budgets per commercial. Consider the commercials aired during the Super Bowl every year. National commercials employ Hollywood film directors, writers, cinematographers, lighting and sound artists, and well known actors or actresses. Successful clients have the financial stability to permit producers to use experimental and original visual techniques and persuasive methods. This is also the type of commercial chosen for Cleo Awards, the Academy Award of television commercial advertising.

The locally produced commercial, on the other hand, features a local client, a small budget, usually a three camera, in-studio production. The commercial is made for local distribution only and is generally produced by the station's advertising sales people. But in many markets, these commercials are the lifeblood for local television

sales and advertising. For local clients, these commercials are the only affordable, far-reaching advertising available and producible. Many of the local clients see these commercials as the opportunity to feature themselves and their children. Automobile dealerships, department stores, and specialty shops appear to make up the majority of these clients.

That local commercials are the scourge of television advertising does not mean that they should be poorly produced and lacking in aesthetics and good taste. Adequate studio equipment should be able to spell some creativity and innovation. One formula to quality commercial production is organization and preproduction. And, should the locally produced commercial still be the lifeblood of the local television station, it behooves beginning professionals, both in academia and in the broadcasting industry, in television production to know and understand where creativity and innovation can be interjected into the production of commercials.

## FLOWCHART AND CHECKLIST

The following pages present in flowchart and checklist fashion the procedure of the preproduction, production, and postproduction stages of the television studio production of the commercial. In this section, the procedures are ordered by the definition of personnel role responsibility at the three stages. Before each entry in the procedure, a checklist box permits notation of the completion of each step as preproduction, production, and postproduction processes proceed.

FLOWCHART AND CHECKLIST FOR THE STUDIO COMMERCIAL PRODUCTION

## PRODUCER (P)

### PREPRODUCTION

- [ ] (1) Contacts/is contacted by the client/agency **11**
- [ ] (2) Researches/reviews marketing problem (marketing analysis form) **11**
- [ ] (3) Writes a commercial script (video script form) **11**
- [ ] (4) Designs the storyboard (storyboard form) **11**
- [ ] (5) Constructs a budget (production budget form) **11**
- [ ] (6) Arranges for the production facility/studio **12**
- [ ] (7) Meets in a preproduction meeting with the client/agency **12**
- [ ] (8) Chooses/meets with the director **13**
- [ ] (9) Creates the production schedule (production schedule form) **14**
- [ ] (10) Holds auditions; supervises casting the commercial with the director **14**
- [ ] (11) Approves set(s) designs; authorizes expenses and purchase requisitions **16**
- [ ] (12) Approves properties list; authorizes expenses and purchase requisitions **17**
- [ ] (13) Approves costume designs; authorizes expenses and purchase requisitions **17**
- [ ] (14) Approves make-up design; authorizes expenses and purchase requisitions **18**
- [ ] (15) Approves audio plot design and music; authorizes expenses and purchase requisitions **23**
- [ ] (16) Approves lighting plot design; authorizes expenses and purchase requisitions **25**
- [ ] (17) Obtains copyright, royalty clearances, and insurance coverage **25**
- [ ] (18) Designs titling, graphics, animation, and photo stills (graphic design form, character generator copy form) **26**

### PRODUCTION

- [ ] (1) Supervises crew/cast calls; meets with director and cast/crew **30**
- [ ] (2) Handles studio/set(s) arrangements **30**
- [ ] (3) Handles crew/cast details **30**
- [ ] (4) Confers with the director on acceptable take(s) **43**
- [ ] (5) Prepares for the production of stills and the product shot (talent release form) **43**
- [ ] (6) Secures talent signatures on the talent release form **44**

### POSTPRODUCTION

- [ ] (1) Arranges postproduction schedule and facilities **46**
- [ ] (2) Supervises postproduction preparation **46**
- [ ] (3) Observes editing with the director; assists editing decisions **47**
- [ ] (4) Reviews the rough edit with the director **47**
- [ ] (5) Previews the rough edit with the client/agency **47**
- [ ] (6) Approves the final edit **48**
- [ ] (7) Turns the master tape over to the client/agency **48**

## DIRECTOR (D)

### PREPRODUCTION

- [ ] (1) Meets with the producer **13**
- [ ] (2) Reviews the script and storyboard **14**
- [ ] (3) Does a script breakdown (script breakdown form) **14**
- [ ] (4) Casts the commercial with the producer **14**
- [ ] (5) Chooses/meets with the set designer; supervises set(s) design(s) **15**
- [ ] (6) Approves set(s) design(s) **16**
- [ ] (7) Chooses/meets with the property master; supervises property selection **16**
- [ ] (8) Approves property plot/list **16**
- [ ] (9) Chooses/meets with the costume master/mistress; supervises costume selection **17**
- [ ] (10) Approves costume designs **17**
- [ ] (11) Chooses/meets with the make-up artist; supervises make-up design **17**
- [ ] (12) Approves make-up designs **18**
- [ ] (13) Designs blocking plots and master script; plans stock shots (blocking plot form) **18**
- [ ] (14) Chooses/meets with camera operators **19**
- [ ] (15) Chooses/meets with the audio director; assigns the audio plot to the audio director **22**
- [ ] (16) Approves the audio plot from the audio director **23**
- [ ] (17) Chooses/meets with the lighting director; assigns the lighting plot **24**

### PRODUCTION

- [ ] (1) Holds production meeting with crew/cast; sets first take blocking time **30**
- [ ] (2) Supervises final studio/set(s) arrangement **30**
- [ ] (3) Supervises equipment set-up/placement **30**
- [ ] (4) Begins blocking; blocks actor(s)/actress(es) *without* lines **35**
- [ ] (5) Rehearses actor(s)/actress(es) *without* lines; walks through blocking **35**
- [ ] (6) Rehearses actor(s)/actress(es) blocking *with* lines **35**
- [ ] (7) Blocks cameras as cast hits spike marks **36**
- [ ] (8) Rehearses both cameras and actor(s)/actress(es) **36**
- [ ] (9) Makes changes and rehearses in studio **38**
- [ ] (10) Retires to the control room; rehearses from the control room **40**
- [ ] (11) Readies shots to camera operators; takes shots to the technical director **40**
- [ ] (12) Calls a break for costuming/make-up **40**
- [ ] (13) Calls for a take(s) or another rehearsal(s) after costuming/make-up **40**
- [ ] (14) Finishes take; stops tape **41**
- [ ] (15) Confers with the producer on acceptable take **43**
- [ ] (16) Produces any required video/stills/shots **43**
- [ ] (17) Announces intentions: take/wrap/next take/strike **44**

### POSTPRODUCTION

If the director edits:
- [ ] (1) Supervises postproduction preparation **46**
- [ ] (2) Reorders/edits master script **46**
- [ ] (3) Prepares postproduction cue sheet form (postproduction cue sheet form) **46**
- [ ] (4) Edits a rough edition **47**
- [ ] (5) Reviews the rough edit with the producer **47**
- [ ] (6) Previews the rough edit with the producer and client/agency **47**
- [ ] (7) Makes the final edit **48**
- [ ] (8) Labels the edited master; removes the record button **48**
- [ ] (9) Turns the master tape over to the producer **48**

If the director does not edit:

### TECHNICAL DIRECTOR/EDITOR (TD/E)

- [ ] (1) Edits from the director's master script, postproduction cue sheet form, and videotape log form under the supervision of the producer and the director **47**
- [ ] (2) Adds music/effects and mixes channels **48**
- [ ] (3) Labels the edited master; removes the record button **48**
- [ ] (4) Turns the master tape over to the producer **48**

- (18) Approves the lighting plot **25**
- (19) Chooses/meets with the technical director **25**
- (20) Chooses/meets with the assistant director and production assistant **26**
- (21) Chooses/meets with the floor director and continuity person **26**
- (22) Chooses/meets with the videotape recorder operator **26**
- (23) Creates/distributes the camera shot lists (camera shot list form) **26**
- (24) Reviews completed set(s) with the set designer **27**
- (25) Checks lighting design and preliminary set(s) lighting **27**
- (26) Reviews properties/demonstrations of properties **27**
- (27) Previews completed costumes **27**
- (28) Views make-up tests **28**

## CAMERA OPERATORS (CO)

- (1) Meet with the director **22**
- (2) Work with the master script **22**

- (1) Meet with the producer/director at crew call or at the production meeting **31**
- (2) Set up/prepare the assigned camera/placement/aim/focus for shading **31**
- (3) Review the shot list, attach the shot list to the camera **32**
- (4) Check intercom connections; announce readiness **34**
- (5) Observe actor(s)/actress(es) blocking and rehearsal by the director **35**
- (6) Are blocked with actor(s)/actress(es); cameras are spiked **36**
- (7) Note details of placement/composition/framing; update shot list **36**
- (8) Rehearse camera with actor(s)/actress(es) **38**
- (9) Update camera/actor(s)/actress(es) changes on the shot list **38**
- (10) Rehearse as called by the director **38**
- (11) Watch for microphone/boom shadows in shots during rehearsals/takes **39**
- (12) Shoot camera for take(s) **41**
- (13) Prepare to block the next unit; strike camera/cables at the director's wrap/strike call **44**

## AUDIO DIRECTOR (A)

- (1) Meets with the director **22**
- (2) Works with the master script/studio set(s) design(s) **23**
- (3) Designs audio coverage and microphone plots (audio plot form) **23**
- (4) Designs sound effects and music needs (effects/music breakdown form) **23**
- (5) Submits the audio plot to the producer and director for approval **23**
- (6) Chooses/meets studio microphone boom grip(s)/operator(s) **23**
- (7) Records/secures prerecorded tracks **24**

- (1) Meets with the producer/director at crew call or at the production meeting **31**
- (2) Meets microphone boom grip(s)/operator(s); begins studio audio equipment set-up **32**
- (3) Runs microphone cables; records studio microphone input(s) **32**
- (4) Patches inputs into audio control board in the control room **32**
- (5) Checks network intercom connections with the director, technical director, and audio control board **32**
- (6) Places the microphone boom grip(s)/operator(s) with microphone boom(s) for the script unit to be blocked **35**
- (7) Makes microphone audio level checks **39**
- (8) Makes studio foldback sound check for microphone boom grip(s)/operator(s) **39**
- (9) Rehearses prerecorded track(s) and sound effects into the studio **39**

- (1) Provides music track(s) and sound effects track(s) (effects/music cue sheet form) **46**
- (2) Assists the producer/director for audio problems/additions to master edit **47**
- (3) Sweetens the final audio track for the edited master **48**

| PREPRODUCTION | PRODUCTION | POSTPRODUCTION |
|---|---|---|
| | (10) Observes actor(s)/actress(es) blocking; directs microphone boom(s) placement with microphone boom grip(s)/operator(s) **39** | |
| | (11) Monitors microphone placement and sound levels during rehearsal(s) **39** | |
| | (12) Makes microphone boom grip(s)/operator(s) changes before the next rehearsal **40** | |
| | (13) Mixes in prerecorded track(s) and sound effects as required by the script **40** | |
| | (14) Monitors actor/actress dialogue and prerecorded track(s) and sound effects during rehearsals **40** | |
| | (15) Records studio dialogue and mixes effects during take(s); monitors for any extraneous studio sounds during take(s) **41** | |
| | (16) Prepares for the next take with a wrap call by the director; disassembles the studio microphone/boom/cable equipment; closes down the control board with a strike call by the director **44** | |

## LIGHTING DIRECTOR (LD)

| PREPRODUCTION | PRODUCTION | POSTPRODUCTION |
|---|---|---|
| (1) Meets with the director **24** | (1) Meets with the producer/director at crew call or at the production meeting **31** | |
| (2) Works with the master script, studio set(s) designs, and property lists **25** | (2) Lights the set(s) and performs lighting check; announces readiness **31** | |
| (3) Designs lighting plots (lighting plot form) **25** | (3) Watches the set(s) and actor(s)/actress(es) during rehearsals for light/shadows **39** | |
| (4) Submits lighting plots to the producer/director for approval **25** | (4) Adjusts lighting/lights after each rehearsal as needed **39** | |
| (5) Lights the completed set(s) **27** | (5) Studies lighting effects over the studio floor monitors; makes final changes **40** | |
| (6) Reviews the light pattern on the studio set(s) with the director **27** | (6) Monitors the set(s) and actor(s)/actress(es) lighting over the control room monitors during take(s) **41** | |
| | (7) Prepares for the next take; strikes the floor lights; turns off set lights at a strike call **44** | |

## CONTINUITY PERSON (CP)

| PREPRODUCTION | PRODUCTION | POSTPRODUCTION |
|---|---|---|
| (1) Meets with the director **26** | (1) Meets with the producer/director for crew call or at the production meeting **31** | |
| (2) Prepares forms for recording continuity notes during rehearsal(s)/take(s) (continuity notes form) **26** | (2) Updates forms for recording continuity details during rehearsal(s)/take(s) **33** | |
| | (3) Records actor(s)/actress(es), set(s), dialogue, and costume details from take-to-take **42** | |
| | (4) Releases actor(s)/actress(es) after take(s)/wrap **42** | |
| | (5) Establishes previous take details before succeeding take(s) **43** | |
| | (6) Reorders continuity forms with studio wrap/strike **44** | |

## PRODUCTION ASSISTANT (PA)

| PREPRODUCTION | PRODUCTION | POSTPRODUCTION |
|---|---|---|
| (1) Meets with the director **26** | (1) Meets with the producer/director for crew call or at the production meeting **31** | |
| (2) Secures titling/screen text list from the producer; enters character generator copy (character generator copy form) **26** | (2) Checks intercom network connections **33** | |
| | (3) Enters/records character generator copy for script lines to be produced during each production session **33** | |
| | (4) Changes the relevant information of each slate screen for succeeding takes **41** | |
| | (5) Turns the character generator off with wrap/strike call **44** | |

## MICROPHONE BOOM GRIP(S)/OPERATOR(S) (MBG/O)

- [ ] (1) Meet with the audio director 23
- [ ] (2) Work with the master script 24
- [ ] (3) Design sound pickup plots (audio pickup plot form) 24
- [ ] (4) Prepare studio microphone(s)/holder(s)/extension(s) 24

- [ ] (1) Meet with the producer/director at crew call or at the production meeting 31
- [ ] (2) Meet with the audio director; begin studio audio equipment set-up 32
- [ ] (3) Assemble microphone boom(s) with microphone(s); run audio cable(s) 32
- [ ] (4) Take place(s) for preliminary dialogue and sound coverage 36
- [ ] (5) Watch actor/actress blocking for microphone coverage; confer with the audio director 38
- [ ] (6) Begin sound coverage during camera blocking with lines by actor(s)/actress(es) 38
- [ ] (7) Confer with the audio director for improved microphone placement; make changes when grip/operator or microphone/boom shadows appear on-camera 39
- [ ] (8) Note CU/LS framing for placement of microphone(s) for improved audio perspective 39
- [ ] (9) Monitor dialogue with foldback/headsets during rehearsal(s)/take(s) 39
- [ ] (10) Rehearse with actor(s)/actress(es) and cameras 39
- [ ] (11) Make necessary changes/adaptations 39
- [ ] (12) Cover actor(s)/actress(es) during take(s) 41
- [ ] (13) Prepare for next take with wrap call by the director; disassemble studio microphone(s)/boom(s)/cable(s) with wrap/strike call by the director 44

## VIDEO ENGINEER (VEG)

- [ ] (1) Checks cameras with camera operators 27
- [ ] (2) Does preventive maintenance 27
- [ ] (3) Checks teleprompter monitors on cameras 27

- [ ] (1) Uncaps cameras 31
- [ ] (2) Readies studio cameras for shading/videotaping 31
- [ ] (3) Checks video levels of cameras with lighting on the set 31
- [ ] (4) Routes external signal to camera monitors for camera operator use 32
- [ ] (5) Alerts the camera operators/director when ready 32
- [ ] (6) Monitors video level of cameras during videotaping 41
- [ ] (7) Alerts the director to soft focused cameras during videotaping 41
- [ ] (8) Caps cameras with the director's call for a wrap/strike 44

## ASSISTANT DIRECTOR (AD)

- [ ] (1) Meets with the director 26
- [ ] (2) Secures/purchases adhesive colored dots 26
- [ ] (3) Assigns a color to each actor/actress and camera 26

- [ ] (1) Meets with the producer/director for crew call or at the production meeting 30
- [ ] (2) Prepares/assists the director with the master script and paperwork 32
- [ ] (3) Checks intercom network connections for director/self 34
- [ ] (4) Calls/readies actor(s)/actress(es) and cameras by script line(s); spikes actor(s)/actress(es) and cameras 34
- [ ] (5) Follows the master script for/with director; notes changes 34
- [ ] (6) Makes changes to and updates master script for the director 40
- [ ] (7) Follows/keeps track of master script for the director at directing console during control room rehearsal(s)/take(s) 41
- [ ] (8) Double checks director's calls for rehearsal(s)/take(s)/wrap(s)/next unit/strike 42

- [ ] (1) Turns videotape log forms over to the producer 46
- [ ] (2) Assists the director during postproduction 47

| PREPRODUCTION | PRODUCTION | POSTPRODUCTION |
|---|---|---|
| | ☐ (9) Assists the director; keeps track of the process; keeps control room log of takes (videotape log form) **42** | |
| | ☐ (10) Oversees strike in control room; secures master script/videotape log **44** | |

**FLOOR DIRECTOR (F)**

| PREPRODUCTION | PRODUCTION | POSTPRODUCTION |
|---|---|---|
| ☐ (1) Meets with the director **26** | ☐ (1) Meets with the producer/director for crew call or at the production meeting **31** | |
| | ☐ (2) Assumes responsibility for studio/set/cast/crew during studio use **32** | |
| | ☐ (3) Checks intercom network connections **33** | |
| | ☐ (4) Knows what the director needs/requires during rehearsal(s)/take(s) **33** | |
| | ☐ (5) Effects what the director requires while the director is in the control room **42** | |
| | ☐ (6) Informs self/cast/crew of the director's intentions: retake/wrap/next unit/strike **43** | |
| | ☐ (7) Removes old spike marks before new script unit blocking **43** | |
| | ☐ (8) Oversees the director's call(s): take/retake/wrap/next unit/strike **43** | |
| | ☐ (9) Oversees complete strike of the studio **44** | |

If the talent will need/use script prompting:

**TELEPROMPTER OPERATOR (TO)**

| PREPRODUCTION | PRODUCTION | POSTPRODUCTION |
|---|---|---|
| ☐ (1) Reviews the commercial script for talent prompting **27** | ☐ (1) Meets with the producer/director for crew call or at the production meeting **31** | |
| | ☐ (2) Sets-up the teleprompter bed with the script copy **31** | |
| | ☐ (3) Prepares to run the script copy for talent use **31** | |

**TECHNICAL DIRECTOR (TD)**

| PREPRODUCTION | PRODUCTION | POSTPRODUCTION |
|---|---|---|
| ☐ (1) Meets with the director **26** | ☐ (1) Meets with the producer/director for crew call or at the production meeting **31** | |
| ☐ (2) Becomes familiar with the master script **26** | ☐ (2) Checks intercom network connections **33** | |
| | ☐ (3) Notes the assignment of the video signal from the playback videotape deck through the production switcher **33** | |
| | ☐ (4) Cuts from shot to shot as the director calls for each take during rehearsal(s) **38** | |
| | ☐ (5) Responds to the director's calls during videotaped take(s) **41** | |
| | ☐ (6) Assists the director in deciding a take **43** | |
| | ☐ (7) Responds to the director's call: retake/wrap/next unit/strike **43** | |
| | ☐ (8) Shuts down the switcher with a wrap/strike call; secures a copy of the master script **44** | |

**VIDEOTAPE RECORDER OPERATOR (VTRO)**

| PREPRODUCTION | PRODUCTION | POSTPRODUCTION |
|---|---|---|
| ☐ (1) Meets with the director **26** | ☐ (1) Meets with the producer/director for crew call or at the production meeting **31** | ☐ (1) Stripes source/master videotapes with SMPTE time code **47** |
| ☐ (2) Determines videotape stock needs: source tapes and master tape(s) **26** | ☐ (2) Prepares record videotape deck; selects labeled videotape stock **33** | ☐ (2) Turns source tapes and master videotape stock over to the producer **47** |
| ☐ (3) Stripes SMPTE code on stock videotape; codes videotapes; labels videotape stock **27** | ☐ (3) Prepares A-roll and B-roll videotapes **33** | |
| ☐ (4) Checks operation of video recorder(s) **27** | ☐ (4) Chooses videotape playback and record decks; alerts technical director to the choices for switcher routing **33** | |
| ☐ (5) Performs preventive maintenance **27** | ☐ (5) Checks intercom network connections **34** | |
| ☐ (6) Secures master videotape(s) for A-roll and B-roll **27** | ☐ (6) Alerts the director when ready **34** | |

☐ (7) Readies/rolls/records videotape deck with the director's call **41**
☐ (8) Monitors recording/playback videotape machines during videotaping **41**
☐ (9) Stops videotaping with the director's call **41**
☐ (10) Rewinds record deck and source tape when taping session is finished; accurately records the tape content on labels **44**

## TALENT/ACTOR(S)/ACTRESS(ES) (T)

☐ (1) Attends audition (talent audition form) **15**
☐ (2) Learns lines; develops characterization (characterization form) **15**
☐ (3) Attends fittings; checks costumes under lights on-camera **18**
☐ (4) Tries make-up; tests make-up under lights on-camera **19**
☐ (5) Parades costumes before the director for approval **27**

☐ (1) Meets with the producer/director for cast call or the production meeting **30**
☐ (2) Meets the director's schedule for lighting stand-in needs **34**
☐ (3) Follows the director's instructions for blocking/rehearsal/action/cut/freeze/take **34**
☐ (4) Meets with the make-up artist at the studio break for make-up preparation; begins make-up application **40**
☐ (5) Meets with the costume master/mistress for costume distribution **40**
☐ (6) Begins costuming **40**
☐ (7) Makes self available or out of the way during steps in rehearsal(s)/take(s) **40**
☐ (8) Checks with the continuity person between take(s) **43**
☐ (9) Makes frequent checks with the costume master/mistress and make-up artist to review, repair, and touch up **43**
☐ (10) Removes costumes/make-up and cleans up with strike call **44**

## SET DESIGNER (SD)

☐ (1) Meets with the director **15**
☐ (2) Studies the script/storyboard **15**
☐ (3) Designs the studio set(s) (set design form) **15**
☐ (4) Submits preliminary designs to the producer/director for approval **15**
☐ (5) Meets with the properties master/mistress to review set(s) designs **16**
☐ (6) Begins set(s) construction **18**
☐ (7) Completes set(s) construction and reviews set(s) with the director **27**

☐ (1) Meets with the producer/director for cast call or at the production meeting **30**
☐ (2) Oversees the set decoration/dressing for each set **31**

## PROPERTIES MASTER/MISTRESS (PM)

☐ (1) Meets with the director **16**
☐ (2) Meets with the set designer; reviews the set(s) designs **16**
☐ (3) Studies the script/storyboard **16**
☐ (4) Creates a properties breakdown form (properties breakdown form) **16**
☐ (5) Submits the properties form to the producer/director for approval **16**
☐ (6) Begins acquisition of needed properties **17**
☐ (7) Reviews and demonstrates properties with the director **27**

☐ (1) Meets with the producer/director for cast call or at the production meeting **30**
☐ (2) Completes set(s) decoration/dressing with set/hand/action properties **31**
☐ (3) Stands by to handle action properties before and after rehearsal(s)/take(s) **35**
☐ (4) Replaces/replenishes properties consumed/used during rehearsal(s)/take(s) **43**
☐ (5) Replaces/alters properties to the changing needs of the commercial **43**
☐ (6) Collects/cleans properties after take(s)/strike **44**

## COSTUME MASTER/MISTRESS (CM)

☐ (1) Meets with the director **17**
☐ (2) Studies the script/storyboard **17**
☐ (3) Designs costumes (costume design form) **17**
☐ (4) Submits costume design(s) to the producer/director for approval **17**

☐ (1) Meets with the producer/director for cast call or at the production meeting **30**
☐ (2) Meets with actor(s)/actress(es); distributes costumes **40**
☐ (3) Assists actor(s)/actress(es) getting into costumes **40**

| PREPRODUCTION | PRODUCTION | POSTPRODUCTION |
|---|---|---|
| ☐ (5) Acquires/creates/builds costumes **18** | ☐ (4) Watches rehearsal(s)/take(s) for use/abuse/needed repair to costumes **43** | |
| ☐ (6) Calls fittings with actor(s)/actress(es); supervises camera/lighting tests **18** | ☐ (5) Makes adjustments/repairs to costumes **43** | |
| ☐ (7) Parades final costumes before the director **27** | ☐ (6) Retrieves/repairs/cleans costumes after rehearsal(s)/take(s)/wrap(s)/strike **44** | |

MAKE-UP ARTIST (MA)

| PREPRODUCTION | PRODUCTION | POSTPRODUCTION |
|---|---|---|
| ☐ (1) Meets with the director **17** | ☐ (1) Meets with the producer/director for cast call or at the production meeting **31** | |
| ☐ (2) Studies the script/storyboard/characterization **17** | ☐ (2) Meets with actor(s)/actress(es) for last minute instructions/changes in make-up **40** | |
| ☐ (3) Designs make-up (make-up design form) **17** | ☐ (3) Proceeds with make-up application **40** | |
| ☐ (4) Submits make-up design(s) to the producer/director for approval **17** | ☐ (4) Checks actor(s)/actress(es) before each take **40** | |
| ☐ (5) Secures make-up supplies **18** | ☐ (5) Watches take(s) for make-up needs/changes/repair **40** | |
| ☐ (6) Tests make-up on actor(s)/actress(es) under lights on-camera **27** | ☐ (6) Observes studio monitors for additional make-up needs/changes/repair **41** | |
| ☐ (7) Reviews make-up design(s); test(s) for the director **28** | ☐ (7) Knows the rehearsal/take schedule for actor(s)/actress(es); assists in readying actor(s)/actress(es) for subsequent rehearsal/take demands **43** | |
| | ☐ (8) Assists actor(s)/actress(es) in removing make-up; assists cleaning up with strike call **44** | |

# Chapter 2

# Processes of Television Commercial Production

## INTRODUCTION

Perhaps the most structured television studio genre is that of the television commercial. The television commercial is exactly timed down to the second. In television newscast and television studio drama production, which are also fully scripted productions, there exists in each room for creativity and innovation; both genres can deviate somewhat from the script. In the television commercial, the production session and the script are constrained by the time length of the commercial. With adequate and thorough preproduction, the commercial script is orchestrated and preplanned to the second; little or no deviation can be allowed. In fact, frequently during commercial production, a script unit is redone again and again until it achieves the preplanned time limit for that unit. This becomes crucial to commercial postproduction editing ease.

Adequate and thorough production of the television commercial is the function of equally adequate and thorough production crew organization and role task accomplishment. These goals are accomplished through the processes of preproduction, production, and postproduction.

## THE PREPRODUCTION PROCESS

Television commercial production differs little from studio drama production; the only real difference may be in the length of the script and the final video product. The teleplay script length is calculated by production units (many of them); the television commercial script is calculated by lines of copy and storyboard images.

## • Personnel

The personnel required for the studio production of the television commercial takes its point of reference from the production of television studio drama. The operation of the production studio requires a minimum number of personnel without which production suffers. Admittedly, in some network affiliate studio operations, a director functions as a technical director and assistant director as well. One person could also function as the set designer and the properties master/mistress, as costume master/mistress and make-up artist, or talent can serve as his or her own costumer and make-up artist. In this text, separate roles will be considered on the premise that once roles are described, role tasks can be integrated and collapsed into fewer roles if the number of available crew is limited. On the other hand, in academia, there is often the need for more roles to accommodate more students. This text attempts to account for all personnel demands.

*Producer (P)* The producer in television commercial production begins the process of preproduction. The producer may be hired by an advertising agency, may be a member of a station's advertising sales department, or may be a freelancer looking for a client. A principal role for a producer, especially a freelancer, is writing the commercial script. The preproduction responsibilities for the producer include creating and keeping the budget, choosing a director, and handling most details from obtaining clearances and insurance coverage to creating titling and graphics for the commercial. A producer must be organized, thorough, and detail-oriented.

*Director (D)* The director plays an equally strong preproduction role as well as the main production role. All

**FIGURE 2–1**
**Production personnel organizational chart.** This organizational chart places all production personnel in their respective supervisory and subordinate positions. Personnel are also classified by primary role category and production facility placement.

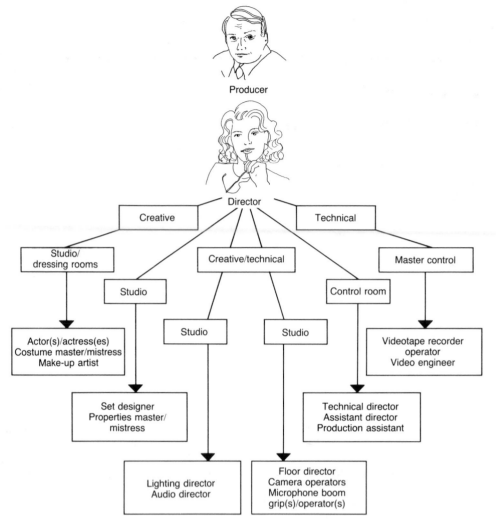

details of studio production preparation fall to the director. Almost all creative and artistic elements of the commercial are the responsibility of the director. A director must have experience with directing actors/actresses and experience in television studio production.

***Camera operators (CO)*** The studio camera operators have minimum but important preproduction responsibilities. They have to become familiar with the script and with their respective camera shot lists.

***Lighting director (LD)*** The lighting director's preproduction responsibilities are creative, at the point of designing the lighting for the commercial, and technical, when producing the lighting design with the studio light instruments.

***Audio director (A)*** The audio director has preproduction responsibilities ranging from the aesthetic, which includes designing the audio plot and effects, to the technical, which entails physically placing cable runs and microphones for the studio production.

***Microphone boom grip(s)/operator(s) (MBG/O)*** The microphone boom grip(s)/operator(s) spend preproduction time working out studio audio pickup design. By

becoming familiar with camera placement and actor/actress blocking, a reasonable idea of microphone boom placement can be gained.

***Assistant director (AD)*** The assistant director works closely with the director in preproduction, but has minimal preproduction responsibilities.

***Floor director (F)*** The floor director has no preproduction responsibility aside from general studio preparation and script study.

***Technical director (TD)*** The technical director's preproduction responsibilities rest in knowing what visual techniques and special effects may be required from the switcher.

***Videotape recorder operator (VTRO)*** The videotape recorder operator has the preproduction responsibility to determine the videotape stock necessary for production and postproduction.

***Video engineer (VEG)*** The video engineer does not have preproduction responsibilities, but is essential to the studio during production.

*Production assistant (PA)* The production assistant's primary preproduction responsibility is to enter and record all character generator copy for production, which includes slate, titling, and any other required screen text.

*Teleprompter operator (TO)* In studio production of television commercials, it is not uncommon for the talent to need script copy prompting. In that case, a teleprompter operator will be added to the studio production crew.

*Continuity person (CP)* The continuity person prepares the system of recording continuity details in the studio.

*Set designer (SD)* The set designer is responsible for designing the studio set(s) and overseeing construction of the set(s).

*Properties master/mistress (PM)* The properties master/mistress is responsible for determining and securing all set, hand, and action properties needed for production. This includes the presentation of the client's product for videotaping (e.g., food preparation or getting an automobile into a studio).

*Costume master/mistress (CM)* The costume master/mistress is responsible for determining costume needs and for obtaining the necessary costumes and costume changes required by the talent.

*Make-up artist (MA)* The make-up artist is responsible for determining the make-up requirements for the talent, designing make-up, and obtaining the necessary make-up supplies.

*Talent (T)* The actor(s)/actress(es) for the commercial are responsible for working on their characterization and learning lines for the production. They are also required to meet make-up design requirements, costume measurement, and costume fittings. On-camera tests will be scheduled to test make-up and costumes under studio lights and on the video screen.

## • Preproduction Stages

*Contacting a client or being contacted by a client or an agency (P 1)* The producer begins preproduction by searching out a client or by being contacted by a client or agency. In academia, a student producer may go into the community and find a client for whom to create a commercial. The producer is the starting point for the preproduction of a television studio commercial.

*Researching and reviewing the marketing problem for the commercial (P 2)* The basic purpose of a commercial is to solve a marketing problem. Marketing problems can be as simple as getting a product or service known to the public to the challenge of positioning a product or service in the marketplace to win a target audience. The best place for a producer to begin is to set narrow goals and objectives for the commercial and to define the target audience. Once goals and objectives are set, a production statement can be created. Together with the goals and objectives, a target audience, and the production statement, video production values and techniques can be chosen to orchestrate the final commercial to the goals, audience, and statement.

*Writing the script (P 3)* Writing the script is crucial because it involves the choice of words, phrases, logic, and emotion that serve, among other perceptual elements of the medium and its technology, to hook the audience and influence them favorably toward the product or service being advertised. Because the audience has only a single pass at hearing the copy, the producer writes for the ear (not the eye, as done in print advertising). Writing for the ear is the most common form of human communication: conversation. A good commercial scriptwriter writes as one speaks—conversationally, often in incomplete sentences and just as often with only a one word sentence.

There is a proverbial defense for writing the voice copy before imaging the commercial. One picture, as the proverb goes, is worth a thousand words. There will be less visual imaging than there will be audio copy. (See the video script form.)

*Designing the storyboard (P 4)* The storyboard as a graphic imaging tool probably has its origin in commercial production. There is no other video genre so refined or so orchestrated that the whole commercial concept is designed, sketched or photographed (called a photoboard), and presented to the client for approval before production begins. The thoroughness and neatness of a storyboard is at the heart of the commercial approval and preproduction process.

The commercial storyboard has no standardized form. It can be constructed with small aspect ratio frames on an 8-1/2 × 11″ piece of paper or with large aspect ratio frames each equal to the size of a sheet of paper. All storyboards should include the imaging of the major video cuts to be contained in the final edited commercial and should be coordinated with the voice-over or dialogue copy that will be heard while that video frame is being seen by the audience. The aspect ratio frames should be drawn or photographed exactly as they will be seen in the commercial. This includes framing details, set details, talent, and the colors of clothing and objects within the frame.

The completed storyboard with the script and proposed music or sound effects is then presented by the producer to the client for changes and approval. (See the video script/storyboard form.)

*Constructing the budget (P 5)* This step is a tedious but important one for the producer. Before any further commitments are made in the production of the commercial, a budget must be prepared by the producer and approved by the client. The client will want to know the total cost of the production of the commercial. The budget must account for all of the elements to be financed as the commercial goes from preproduction to postproduction.

There are many model budget forms available to the producer as a guide to creating an estimated budget. Any budget form can serve as a reminder of the many elements to be considered in estimating the cost of commercial production. (See the production budget form.)

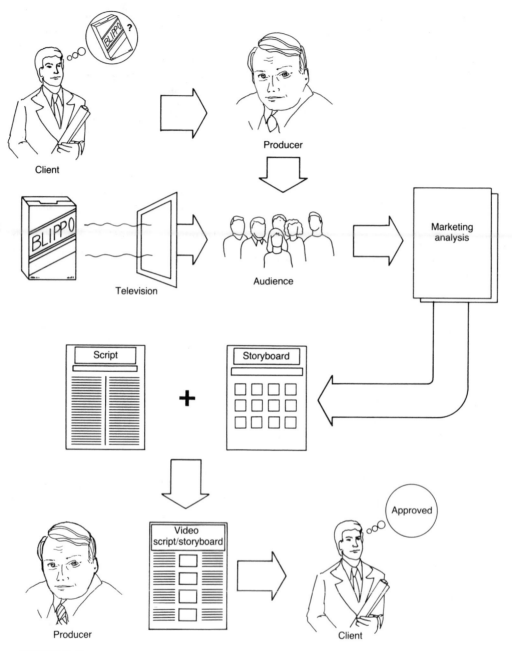

**FIGURE 2–2**
**Client/producer beginning relationship.** The commercial begins with a client and moves to production with client approval. Intermediate stages depend on the creative design of the producer.

***Arranging with a production facility or studio (P 6)***
The producer is responsible for arranging a studio production facility in which to videotape the commercial. Contacting such a facility means obtaining the rate card from that facility for production expenses incurred in the studio. These would be important expenses to cover in a proposed budget. Special needs of the commercial will have to be taken into consideration (e.g., bringing an automobile into the studio or constructing an above-the-ground swimming pool in the studio).

***Meeting with the client or agency (P 7)*** Once the budget is complete and a production facility chosen, the

producer meets with the client or the agency. During this meeting, the script and storyboard are reviewed and approved and the budget is presented and approved. This meeting is very important. At this meeting, wording and imaging for the commercial must be made clear and explicit approval obtained from the client or agency. Approval of the script, storyboard, and expenses is critical to proceeding with the production of the commercial. Proposed music for the commercial is played for approval also. Planned graphics and animation (indicated on the storyboard) are reviewed and approved. Projected cost figures in the budget have to be made clear and certain and then approved. Approval of some graphics (e.g.,

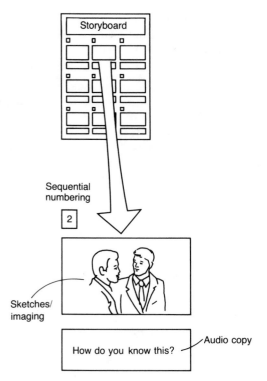

**FIGURE 2-3**
**The storyboard form.** The producer uses the storyboard form to image the draft of the commercial script. This form provides more frames than the client needs for approval.

computer graphics) and animation may entail contracting out the artwork and production to outside agencies.

***Choosing and meeting with a director (P 8)*** With the approval of the budget by the client, the producer chooses and makes a commitment to a director. The director should have complete creative control over all remaining production aspects of the commercial. The director has an obligation to the client and to the approved storyboard and script, but the creative interpretation of the storyboard and script should be that of the director. The director should keep the producer fully informed on the development of and any change to the storyboard or script.

The choice of director by a producer is a critical one. A director expects complete production control and, sometimes, creative control. The producer has chosen the director on the basis of production skills and creative abilities. The first step for the director is to meet with the producer. At this meeting, script, storyboard, and other production values are discussed.

***Meeting with the producer (D 1)*** With the appointment of a director, the producer and director should meet. This meeting affords the producer and director the opportunity to exchange production values and design concepts. The director should receive a copy of the script, storyboard, and budget during this meeting.

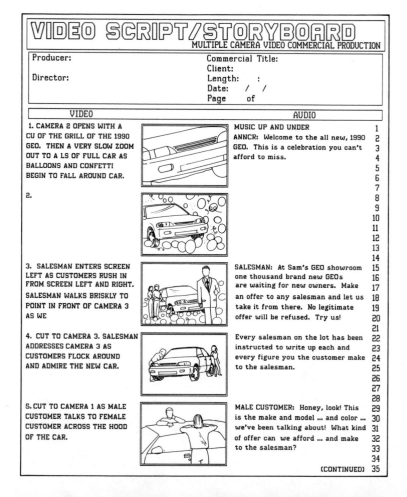

**FIGURE 2-4**
**The script/storyboard.** The script/storyboard presented to the client for approval is the combination form. This form combines the script and storyboard form. Selected aspect ratio frames of major screen image changes are recorded on this form from the producer's storyboard.

***Reviewing the script and storyboard (D 2)***   The director reviews the script and storyboard for the commercial. Although the producer wrote and designed the script and storyboard, which were then approved by the client, there is still a lot of potential creativity and innovation available to the director. The producer has chosen the director for creative interpretation and visual production values.

***Doing a script and storyboard breakdown (D 3)***   The director begins preproduction tasks by doing a script breakdown. The breakdown is an analysis of the script and the storyboard by set(s), talent, time of day, and interior or exterior setting. (See the script breakdown form.)

Television commercial scripts have to be reorganized into shooting units usually by script lines to facilitate studio production. These line units are created to make multiple camera movement and audio microphone pickup manageable in the studio. The director also needs to be sensitive to the dramatic flow of the script in choosing production units as well as to anticipate continuity problems in editing these units in postproduction. The script breakdown will organize the commercial into shooting units and the order in which these units will be produced in the studio.

***Creating the production schedule (P 9)***   As soon as the budget has been approved, a production facility secured, and a director chosen, the producer creates the production schedule. The production schedule commits to a production session with the studio facility, the cast, and the crew. Crew and cast calls are set with the production schedule. (See the production schedule form.)

***Holding auditions and supervising the casting of the commercial with the director (P 10)***   The producer issues a casting call for prospective actor(s) and actress(es). The producer should defer casting choices to the director as part of the director's creative control of the production. It is best when both the producer and director agree on cast choices. Generally, the producer takes a supervisory role in casting.

In addition to the demands of the script copy and the goals and objectives of the commercial, the profile of the target audience provides guidelines for choosing the right talent. Prospective talent for commercials can be reached usually by advertising open casting calls and having all actors and actresses responding read for the part(s).

***Casting the commercial script (D 4)***   Because one of the primary qualities of a director is creativity, the director should take a leading role in choosing the cast. Choosing the cast is one way in which a director begins to express creativity and innovation. It is preferred that both the producer and director agree on a final cast; final casting should be made the mutual choice of the producer and

**FIGURE 2–5**
**The production schedule.** The producer sets a production schedule for the production session.

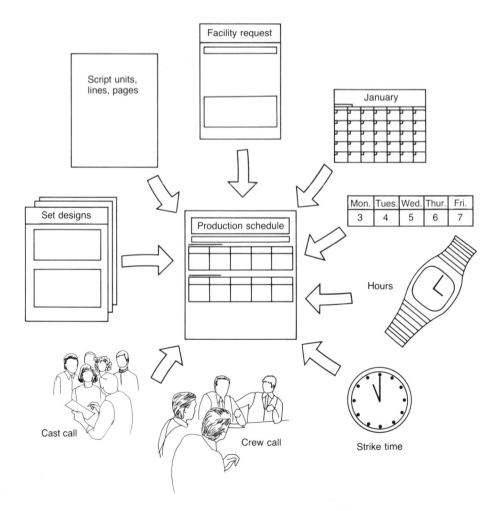

director. In an actual expense venture, costs incurred with union organizations and actors and actresses equity have to be weighed. In academia, relations with theater departments and acting majors are also a consideration. In some amateur productions, actors and actresses can be attracted to the cast for the opportunity to have a television commercial on their resumé/vita and a portfolio videotape. (See the talent audition form.)

***Auditioning for roles in the commercial (T 1)***
Prospective talent audition for roles in the commercial. Talent should attend an audition expecting to be videotaped. There are many ways to audition for roles in a commercial; they include reading parts of the commercial script for the role best suited to a particular actor or actress, improvising, or doing a characterization.

A prospective actor or actress should attend an audition prepared with a photograph of him- or herself, a resumé or vita, and information of membership in any unions or equity organizations. Auditioning talent will be expected to complete an audition information form. (See the talent audition form.)

***Learning lines and developing characterizations (T 2)***  The actor(s) and actress(es) spend their remaining preproduction time learning their lines and developing the characterization required by their script roles. The actor(s) and actress(es) will not be involved in theater style rehearsals because blocking will not occur until the production gets to the studio. Blocking, so common to theater preproduction, cannot be performed this early in commercial production because it cannot be accomplished without studio cameras and camera operators, sets, and properties. Blocking for actor(s) and actress(es) in commercial production is built into the time and process of television studio production. The only rehearsals that

actor(s) and actress(es) should engage in are sit-down reading sessions, common line learning, and delivery sessions. Blocking is futile before actual studio production time. (See the characterization form.)

***Choosing and meeting with the set designer and supervising the set design(s) (D 5)***  The director chooses a set designer. After the director has chosen a set designer, the director meets with the set designer to elaborate on the production requirements and to assist the set designer in beginning set design responsibilities. It is important that the director and set designer determine deadlines for the completion of the set design(s).

***Meeting with the director (SD 1)***  The set designer meets with the director as soon as possible after being chosen as a first step in preparing the set design(s).

***Studying the script and storyboard (SD 2)***  The set designer must read and study the script. The script will suggest a functional set design that will serve the production. The set designer will have to know of and work within any studio facility restrictions (e.g., height of ceiling, lighting battens, size of the studio, and the presence of a cyclorama).

***Beginning the design of the set(s) (SD 3)***  The set designer begins as soon as possible to create the design(s) for the set(s), working with front views and bird's eye views on some proportional scale. (See the set design form.)

***Submitting the design(s) for the studio set(s) to the producer and director for approval (SD 4)***  The set designer will have to submit the preliminary design(s) for the set(s) for approval by the producer and director.

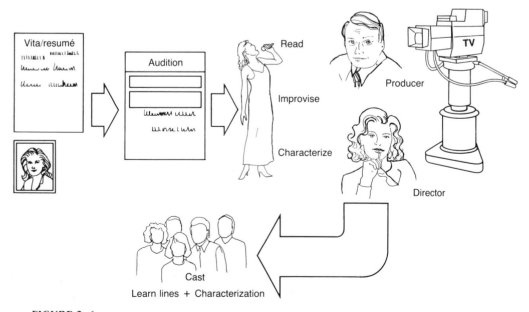

**FIGURE 2–6**
**Talent audition session.** Prospective talent attend an audition with a vita/resumé and photograph and prepare to be videotaped. After being chosen, learning the lines and developing the character are the remaining preproduction responsibilities of the talent.

***Approving the set design(s) (D 6)***   The director makes changes, offers suggestions, and approves the preliminary design(s) of the set(s) before the set designer can proceed. The director requires the set design(s) before the director can begin the master script.

***Approving the set design(s) and authorizing expenses and purchase requisitions (P 11)***   The producer reviews the set design(s), makes changes or suggestions, approves them, and authorizes needed expenses and purchase requisitions for set design and construction.

***Choosing and meeting with a properties master/ mistress and supervising selection of properties (D 7)***   The director chooses and then meets with a properties master/mistress. The properties master/mistress works closely with the set designer. The director shares the aesthetic expectations and plans for the production and sets deadlines for the acquisition of properties.

***Meeting with the director (PM 1)***   The properties master/mistress meets with the director soon after being chosen by the director. The properties master/mistress needs the input of the director in the listing, selection, and acquisition of properties for the commercial.

***Meeting with the properties master/mistress (SD 5)***   Once the set designer has the approval from the producer and director on set design(s), the set designer meets with the properties master/mistress before the properties master/mistress begins the listing and design of properties.

***Meeting with the set designer (PM 2)***   The properties master/mistress meets with the set designer with the approved set design(s). The set design(s) will assist the properties master/mistress in evaluating the set properties for the commercial.

***Studying the script and storyboard (PM 3)***   With the set design(s) in hand, the properties master/mistress studies the script and storyboard for needed properties.

***Creating a property plot or list (PM 4)***   The properties master/mistress creates an exhaustive property plot or list of needed properties from the script. The plot lists all of the needed set properties, action properties, and hand properties. The properties master/mistress notes what properties will require special insurance coverage during production. (See the properties breakdown form.)

***Submitting the properties plot or list to the producer and director for approval (PM 5)***   The properties master/mistress submits the completed properties list to the producer and director for approval.

***Approving the properties list (D 8)***   The director reviews, makes any changes necessary, and approves the properties list for the production.

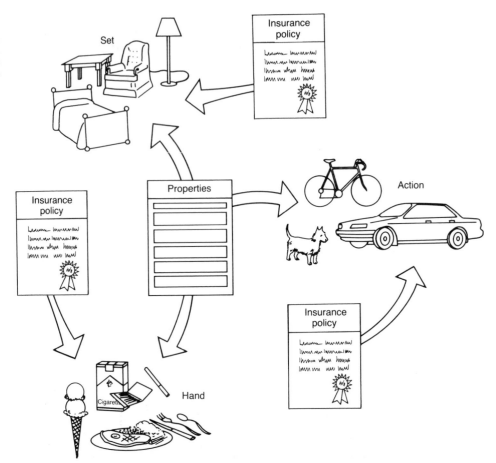

**FIGURE 2–7**
**Properties.** Commercial production makes use of all types of properties. The product being advertised is a property. Properties are classified as hand, set, or action.

*Approving the properties list and authorizing expenses and purchase requisitions (P 12)*   The producer reviews, makes suggestions and changes, and approves the properties list. The producer then authorizes expenses and purchase requisitions for properties acquisition. It is the producer's responsibility to pay special attention to any insurance coverage required for the use of certain properties.

*Beginning the acquisition of properties on the properties list (PM 6)*   The properties master/mistress begins acquiring the properties listed on the properties list as approved by the producer and director. Acquisition may require renting, purchasing, or borrowing properties. Borrowing properties may require insurance coverage.

*Choosing and meeting with the costume master/mistress and supervising costume selection or design and creation (D 9)*   The director chooses a costume master/mistress to oversee the costuming. The director shares production expectations and planning as well as the commercial's interpretation and characterization as an assist in the design or selection of costuming or the creation and building of costumes for the production.

*Meeting with the director (CM 1)*   The costume master/mistress meets with the director before proceeding with costume design. The director is the resource person with insight into the aesthetic approach and interpretation of the script. An aesthetic approach and script interpretation will translate into costume choices of the correct period and color schemes.

*Studying the script and storyboard (CM 2)*   The costume designer studies the script and storyboard as the next step to designing costumes. A script breakdown of costume needs and characters should result from studying the script.

*Designing the costumes (CM 3)*   The costume master/mistress begins to design costumes for individual characters and changes of costumes for characters within the commercial production. Costume design forms that suggest elements of costume design to be considered should be used. (See the costume design form.)

*Submitting costume designs to the producer and director for approval (CM 4)*   The costume master/mistress submits the costume design plots to the producer and director for approval.

*Approving the costume design(s) (D 10)*   The director reviews, makes any necessary changes, and approves the costume design(s).

*Approving costume design(s) and authorizing expenses and purchase requisitions (P 13)*   The producer reviews the costume design(s), makes any necessary changes, and approves the design(s). The producer authorizes any expenses and purchase requisitions for costume construction materials or for the rental of costumes. If talent can use personal clothes, then they should be reimbursed for that use.

*Choosing and meeting with the make-up artist and supervising make-up design (D 11)*   The director continues with preproduction details on the cast side of the production by appointing and meeting with a make-up artist. The director shares production requirements for make-up with the make-up artist as well as the production expectations, plans, interpretations, and characterization of roles.

*Meeting with the director (MA 1)*   The make-up artist meets with the director as the first step in the design of make-up for the cast.

*Studying the script, storyboard, and talent characterization (MA 2)*   The make-up artist begins preproduction work by studying the script and the characterization of the roles in the commercial in preparing to begin make-up design for the cast.

*Designing make-up (MA 3)*   The make-up artist uses the make-up design and preparation forms and creates make-up design(s) and design changes for the cast. Part of designing make-up includes determining make-up products and supplies to be purchased. (See the make-up design form.)

*Submitting the make-up design(s) to the producer and director for approval (MA 4)*   The make-up artist

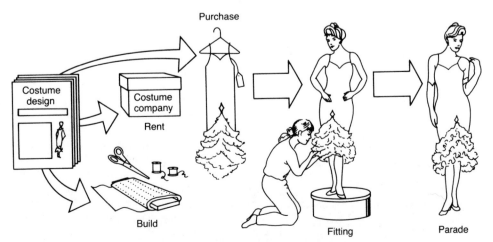

Purchase

Costume design

Costume company

Rent

Build

Fitting

Parade

**FIGURE 2–8**
**Costuming.** Costuming the commercial production involves the design, acquisition or building of costumes, fitting costumes, and parading them under the lights and on-camera before production.

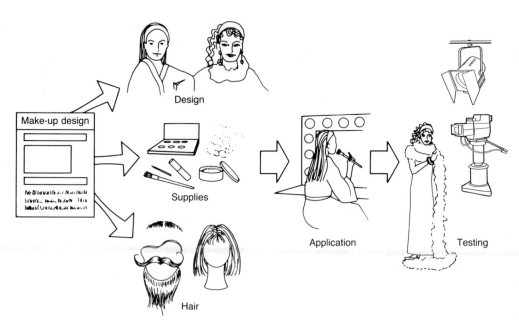

**FIGURE 2–9**
**Make-up.** Make-up preparation involves designing, obtaining supplies, applying, and testing under lights and on-camera.

submits all make-up design(s) to the producer and director for final approval.

***Approving make-up design(s) (D 12)*** The director reviews the make-up design(s), makes any additions or changes, and approves the design(s).

***Approving make-up design(s) and authorizing expenses and purchase requisitions for make-up supplies (P 14)*** The producer reviews all make-up design(s), makes necessary changes, and approves the design(s). The producer authorizes any necessary expenses and purchase requisitions for make-up supplies.

***Securing required make-up supplies (MA 5)*** With the approval of the producer and director, the make-up artist begins purchasing or collecting the necessary make-up supplies and materials.

***Acquiring and building the approved costumes (CM 5)*** With costume design approval from the producer and director, the costume master/mistress begins acquiring by rental or loan, or by building the costumes by purchasing needed materials, renting sewing equipment if necessary, and putting together a crew for working on the costumes.

***Calling fitting appointments with cast when necessary and testing costumes under the lights on-camera (CM 6)*** The costume master/mistress calls cast fittings for costumes occasionally during costume construction. At some point during construction, costumes will have to be judged under studio lights and over camera monitors for their true representation in video.

***Attending scheduled fittings, costume checks, and dress rehearsals on-camera under studio lights (T 3)*** The cast will have to be available for all scheduled fittings for costumes. The cast will also have to parade costumes in dress rehearsals under studio lights and over camera monitors before production begins.

***Creating the master script for the production and designing the storyboard (D 13)*** With approval of the set design(s), the director's major preproduction task and pivotal preproduction document, the master script, can be begun. The master script is the copy of the script from which the commercial will be directed and from which most other preproduction stages will be taken (e.g., the shot list and lighting design).

The director begins master script preparation by numbering all of the lines of the script, including stage direction lines. This is most conveniently done by placing a number, usually every tenth number, opposite the respective line. This numbering provides reference units for subsequent production.

The master script contains a combination of two elements. The first element for every page is a left-hand column containing the actual script. The second element is a threefold right-hand column. The right-hand column is a series of aspect ratio frames in tri-column form corresponding to each of the three studio cameras. The director may have to prepare the master script pages from the original script by photocopying the script into a left-hand column for ease and convenience in designing the master script. In the right-hand column, the director will sketch, in frames for each camera, any change to be effected during production. Each storyboard sketch will be drawn next to the script to which it refers. These storyboard frames are an elaboration on the storyboard frames provided by the producer. The producer's storyboard was not intended to exhaust all possible camera shots, whereas the director's master script should contain sketches of all camera shots.

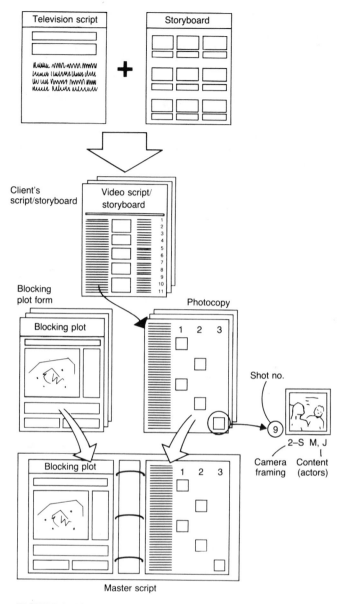

**FIGURE 2–10**
**The development of the director's master script.** The director's master script is an evolution from the producer's script/storyboard, the client's approved script/storyboard, and the director's own creative interpretation and design. The director's blocking plot is the final storyboard from which the commercial will be directed.

Before beginning to decide on and sketch the storyboard frames, the director designs blocking plots containing bird's eye views of all of the set design(s). The director sketches the proposed blocking of all set properties, actor(s)/actress(es), and their movement on the bird's eye view of the set. The three studio cameras must be placed in position on the set(s). The director then blocks actor(s), actress(es), and cameras through the script by using circles with a character's initial within the circle. The circles are repeated at each character's rest position. Arrows connect the character's circles as a character moves around the set.

When a bird's eye drawing becomes too complicated with circles and arrows, blocking should be continued on another blocking plot form. It is recommended that the director make a few copies of the set design(s) at the start of the blocking stage.

To prepare a master script, the director should place all script and storyboard columns on the right-hand side of a loose-leaf notebook page and a page of the blocking plot on the left-hand side facing the script and storyboard for which the blocking corresponds. Blocking plots should be photocopied for any other pages of script using the same blocking and placed opposite their continuing script dialogue.

With the blocking plots completed, the director returns to the script pages. Each planned camera change will be storyboarded. A simple sketch, often just stick figures with circles for heads, is created in successive storyboard frames corresponding to the script dialogue or stage directions. The director then decides for each storyboard frame (1) camera (by number—1, 2, 3—and position—left to right) placement, (2) the framing of the shot (noted by XCU, CU, MS, LS, XLS), (3) any camera movement during the shot (e.g., zooming, arc, trucking, dollying, or pedestaling), and (4) the content of the framing (e.g., "M" for the character Mary). Each storyboard frame is then numbered (across the three cameras) consecutively. Some directors will skip every tenth number, leaving a number to insert additional storyboard frames at a later date, perhaps during studio blocking. Each storyboard frame is then related by a horizontal line drawn to the point in the script dialogue or stage directions where the camera shot is to be changed. A slash is placed in the script dialogue or stage directions at the exact point of camera change.

Directors are reminded that in dialogue cutting in television production, camera changes (switcher cuts, for example) are taken *before* the last word of the line a character is speaking at an anticipated camera change. This is done for aesthetic reasons, in order to have the character responding in dialogue on camera as the characters continue the dialogue. Vertical line arrows can link successive storyboard frames to long passages of script for which a particular shot is to be held.

The director also makes note of any exterior stock shots that may be required for the commercial. These shots will have to be taken sometime during production or drawn from a videotape library.

***Building of the approved set design(s) begins (SD 6)***
The set designer begins construction of the set(s).

***Trying make-up design(s) and testing the design(s) under lights and over the camera monitors (T 4)***
Before studio production, each member of the cast tries the assigned make-up design on and tests the make-up under studio lights and over the camera monitors.

***Choosing and meeting with the camera operators (D 14)*** The director chooses and meets with the three studio camera operators. Choosing good camera operators is essential to the success of the production. Each camera has to be blocked, and the camera operators learn blocking

**FIGURE 2–11**
**The commercial script as part of the master script.** The approved script is prepared for storyboarding by being photocopied to a left-hand column of what will become the master script storyboard.

| | |
|---|---|
| MUSIC UP AND UNDER | 1 |
| ANNCR: Welcome to the all new, 1990 | 2 |
| GEO. This is a celebration you can't | 3 |
| afford to miss. | 4 |
| | 5 |
| | 6 |
| | 7 |
| | 8 |
| | 9 |
| | 10 |
| | 11 |
| | 12 |
| | 13 |
| | 14 |
| SALESMAN: At Sam's GEO showroom | 15 |
| one thousand brand new GEOs | 16 |
| are waiting for new owners. Make | 17 |
| an offer to any salesman and let us | 18 |
| take it from there. No legitimate | 19 |
| offer will be refused. Try us! | 20 |
| | 21 |
| Every salesman on the lot has been | 22 |
| instructed to write up each and | 23 |
| every figure you the customer make | 24 |
| to the salesman. | 25 |
| | 26 |
| | 27 |
| | 28 |
| MALE CUSTOMER: Honey, look! This | 29 |
| is the make and model ... and color ... | 30 |
| we've been talking about! What kind | 31 |
| of offer can we afford ... and make | 32 |
| to the salesman? | 33 |
| | 34 |
| (CONTINUED) | 35 |

**FIGURE 2–12**
**The director's blocking plot.** The director blocks the cast on a bird's eye view of the set and properties. The director blocks the studio cameras as each cast member is blocked.

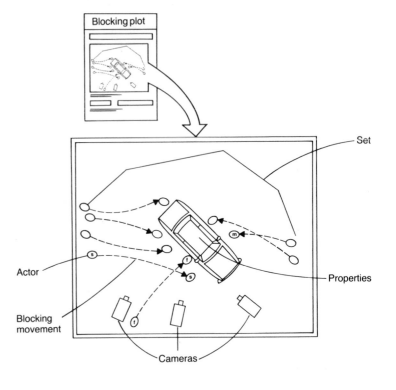

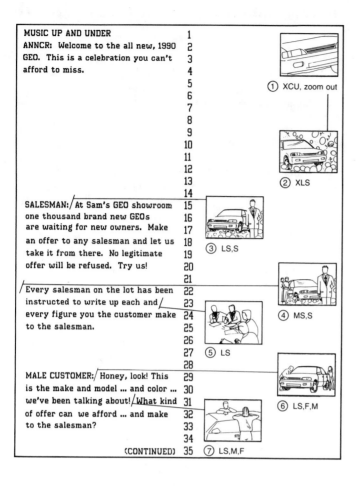

| | |
|---|---|
| MUSIC UP AND UNDER | 1 |
| ANNCR: Welcome to the all new, 1990 | 2 |
| GEO. This is a celebration you can't | 3 |
| afford to miss. | 4 |
| | 5 |
| | 6 |
| | 7 |
| | 8 |
| | 9 |
| | 10 |
| | 11 |
| | 12 |
| | 13 |
| | 14 |
| SALESMAN: / At Sam's GEO showroom | 15 |
| one thousand brand new GEOs | 16 |
| are waiting for new owners. Make | 17 |
| an offer to any salesman and let us | 18 |
| take it from there. No legitimate | 19 |
| offer will be refused. Try us! | 20 |
| | 21 |
| / Every salesman on the lot has been | 22 |
| instructed to write up each and / | 23 |
| every figure you the customer make | 24 |
| to the salesman. | 25 |
| | 26 |
| | 27 |
| | 28 |
| MALE CUSTOMER: / Honey, look! This | 29 |
| is the make and model ... and color ... | 30 |
| we've been talking about! / What kind | 31 |
| of offer can we afford ... and make | 32 |
| to the salesman? | 33 |
| | 34 |
| (CONTINUED) | 35 |

① XCU, zoom out

② XLS

③ LS,S

④ MS,S

⑤ LS

⑥ LS,F,M

⑦ LS,M,F

**FIGURE 2–13**
**Master script storyboard.** Aspect ratio frames are drawn opposite the script lines to which they refer for production imaging.

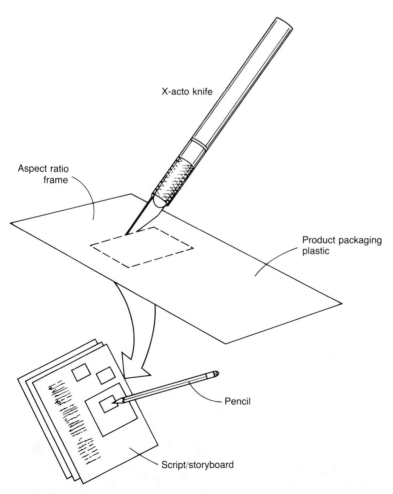

X-acto knife

Aspect ratio frame

Product packaging plastic

Pencil

Script/storyboard

**FIGURE 2–14**
**Aspect ratio template.** A simple template can be made to assist the director in storyboarding the script. Take the heavy plastic packaging from store products. Using an x-acto knife, cut an aspect ratio frame to be used as a template.

in the same way an actor or actress learns blocking. Besides blocking, every camera shot assigned to a camera, the assigned framing, the shot composition, and any camera movement will have to be second nature to the camera operator. Every camera operator has to be a videographer—a photographer with a video camera. The director provides a photocopy of the master script to each camera operator and helps each one prepare for production.

***Meeting with the director (CO 1)***   The camera operators meet with the director to gain some preproduction insight into the aesthetic nature of the commercial as envisioned by the director. They also review expectations of the director during production. Any difficult shots or camera movements expected during studio production are reviewed.

***Working with a copy of the master script (CO 2)*** The camera operators prepare for production by reviewing a copy of the director's master script. The master script

gives the camera operators a better familiarity with the expectations of the director from them and the demands of the script and production.

***Choosing and meeting with the audio director (D 15)***   The director chooses and meets with the audio director for the production. The meeting is a chance for the director to convey sound aesthetics and to share the director's expectations and plans for the production. The director should give the audio director a photocopy of the master script to assist in designing the audio plot.

***Meeting with the director (A 1)***   The audio director meets with the director as a first step in preproduction for sound coverage. The audio director should come away from the meeting with a sense of the director's expectations with regard to the sound aesthetics as well as some idea of the music and sound effects expected for the production. A photocopy of the master script with blocking plots will assist the audio director in designing the audio plot(s).

**FIGURE 2–15**
**The audio plot.** Using the director's blocking plot, the audio director creates the microphone and sound coverage needs of the set.

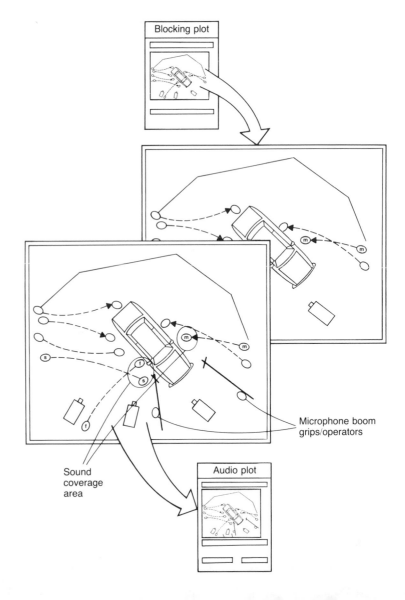

**Working with the master script and studio set designs(s) (A 2)** The audio director works with a copy of the master script and studies the set design(s) as a prelude to designing the audio plot(s).

**Designing the audio plot(s) and microphone plot (A 3)** Using the audio plot form, the audio director designs the audio plot(s). From those plot(s), the audio director determines microphone and cable needs as well as the number of microphone boom grips needed for production. The number of microphone grip personnel and microphone holder equipment can be determined from the distance between actor(s) and actress(es) during dialogue and blocking. The audio director determines at this stage what prerecorded audio will have to be recorded. (See the audio plot form.)

**Designing sound effects and music requirements (A 4)** The audio director designs the sound effects that will be required and determines the music necessary for the sound track. The audio director notes any music or sound effects that require copyright clearance or any other clearance (e.g., synchronization rights). (See the effects/music breakdown form.)

**Submitting the audio plot(s) to the producer and director for approval (A 5)** The audio director submits the audio plot(s) with the hardware, sound effects, and music requirements to the producer and director.

**Approving the audio plot(s) (D 16)** The director reviews the audio plot(s), makes changes or suggestions, and approves the plot(s).

**Approving the audio plot(s) and authorizing expenses and purchase requisitions (P 15)** The producer reviews the audio plot(s), makes necessary changes, and approves the plot(s). The producer authorizes any necessary expenses and purchase requisitions. The producer notes the music requirements and the necessity of obtaining copyright and performance clearances.

**Choosing and meeting with microphone boom grip(s)/operator(s) (A 6)** Once the audio plot(s) is approved, the audio director knows how many studio microphone boom grips/operators will be needed for the production. If there are platform booms, at least two operators will be needed per platform. One operator will operate the boom with the microphone and the other will have to move the platform. If sound coverage is to be done with fishpole booms, one grip will be needed per fishpole. The audio director chooses and meets with the microphone boom grip(s)/operator(s).

**Meeting with the audio director (MBG/O 1)** The microphone boom grip(s)/operator(s) meet with the audio director to review the audio design plot(s) and to plan sound coverage of the studio set(s) and cast. Assignments are made for particular sound coverage units; troublesome shooting units are discussed and solutions suggested.

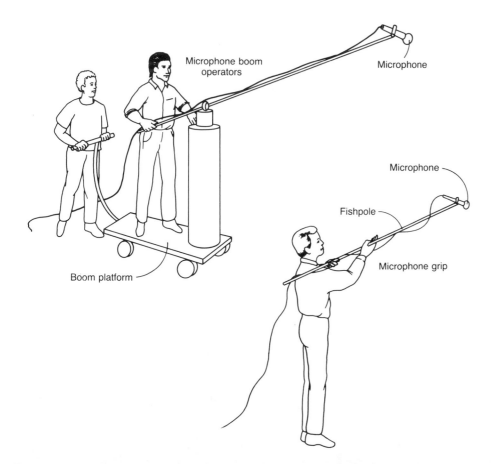

Microphone boom operators

Microphone

Boom platform

Microphone

Fishpole

Microphone grip

**FIGURE 2–16**
**Microphone boom possibilities.** Microphone boom grips and operators might cover the studio set dialogue with a platform boom or a fishpole.

***Working with the master script and audio plot design(s) (MBG/O 2)*** The microphone boom grip(s)/operator(s) spend preproduction time studying the master script and audio plot design(s). The more details they are familiar with, the better prepared they will be for production. Some professionals think that studio microphone boom grip(s)/operator(s) do better at performing sound coverage if they memorize the script for the cast they cover.

***Designing audio pickup plot(s) (MBG/O 3)*** The microphone boom grip(s)/operator(s) design sound pickup requirements. The master script and audio plots suggest positions for grip(s)/operator(s) and microphone holder(s)/extension(s) during production. (See the audio pickup plot form.)

***Preparing studio microphones, microphone holding hardware, and extension(s) (MBG/O 4)*** The microphone boom grip(s)/operator(s) spend some preproduction time preparing the studio microphones to be used in production and obtaining and assembling microphone holding hardware (e.g., fishpoles) and extension(s).

**FIGURE 2–17**
**The lighting plot.** The lighting director uses the director's blocking plot to create the lighting design. Light instruments are placed on the blocking plot and aimed.

***Recording or securing prerecorded audio tracks (A 7)*** The audio director arranges for the recording of any prerecorded tracks that may be needed (e.g., if a portion of the commercial is a voice-over, the voice-over should be prerecorded). On the other hand, some prerecorded tracks may have to be secured from other sources (e.g., from a sound recording production house).

***Choosing and meeting with the lighting director and assigning the lighting plot (D 17)*** The director continues to build the production crew by bringing a lighting director into the crew. The director meets with the lighting director to convey the aesthetic design expectations for the commercial. Lighting is a principal source for mood and environment and is critical in communicating some of the message of the production. The director gives the lighting director a photocopy of the master script for use in learning the camera placement, set(s), time of day, set properties, and blocking of the actor(s) or actress(es). The master script contains all of these details.

***Meeting with the director (LD 1)*** The lighting director meets with the director as the first step in designing

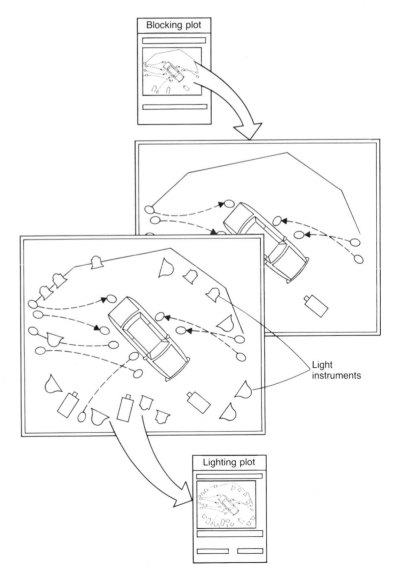

lighting plots. The lighting director needs to have a copy of the master script as preparation to designing the lighting. The lighting director needs to know set design(s) including interior/exterior setting, time of day, set and action property placement, actor and actress blocking, and camera placement. The lighting director also needs to have the director's aesthetic sense for the mood to be conveyed.

**Working with the master script, set design(s), and properties list (LD 2)** The lighting director must study the set design(s) and properties list and create breakdown notes as input to ultimately designing the lighting plots. In some instances, knowing costume colors, make-up, and hair color and styles is helpful to the lighting director in designing the lighting.

**Designing the lighting plots (LD 3)** The lighting director designs the lighting plots for the set(s). (See the lighting plot form.)

**Submitting the set lighting design(s) to the director (LD 4)** The lighting director reviews each lighting design with the producer and director. Final approval of any lighting design cannot come until the sets are complete and in the studio, lighting design(s) are in effect, actor(s) and actress(es) are in place with costumes, and the cameras are on.

**Approving the lighting plot designs and approving expenses and purchase requisitions (P 16)** The producer reviews the lighting plot designs with the lighting director and makes any changes, offers suggestions, and approves the designs. The approval authorizes any expenses and purchase requisitions connected with the designs.

**Approving the set lighting plot design(s) (D 18)** The director reviews and approves the preliminary lighting plot designs. This authorizes the lighting director to begin lighting the studio set(s) as soon as the set designer completes the set(s).

**Obtaining copyright and royalty clearances and insurance coverage (P 17)** The producer is responsible for and can begin to seek legal clearances. Clearances fall into two main areas: (1) music or other copyrighted material being used in the commercial, and (2) claims that may be made as part of the commercial about the product or service being advertised.

An important part of the script and storyboard is the music to be used. The choice of music can make or break the mood or setting that the audio and video portions of the commercial have created. The producer is legally responsible for obtaining all clearances for any copyrighted material, especially music, being used in the commercial. Three options are usually open to a producer: (1) use original music, (2) obtain synchronization rights for some previously published (and copyrighted) music, or (3) use stock/library music for the cost per drop of the music used in the commercial.

The clearances necessary for claims that are made of a product or service demand legal counsel. This counsel should rule that whatever laboratory or scientific data are available for the product or service demonstrate the validity of the claims made in the commercial.

The producer is also responsible for obtaining whatever insurance coverage the production may require. For example, insurance coverage is required by a facility if cigarette, cigar, or pipe smoking is required by the cast; if a real fire is required in a fireplace; or to cover the security of borrowed properties or of moving an automobile in the studio. Insurance may be required if small children not covered under other insurance policies are part of a production and could be injured in the studio.

**Choosing and meeting with the technical director and assigning the shot list breakdown (D 19)** The director chooses and meets with the technical director. The technical director plays an important supportive role to the director. The meeting with the technical director covers the director's aesthetic expectations, as given to the other crew members as well as the director's plans for the production.

**FIGURE 2–18**
**The producer's legal responsibilities.** Before production begins, the producer should secure all necessary rights and clearances, purchase insurance coverage required, and obtain any talent releases necessary.

Producer

*Meeting with the director (TD 1)* The technical director meets with the director as a preliminary step to beginning any preproduction tasks.

*Becoming familiar with the master script (TD 2)* The technical director becomes familiar with the master script as an introduction to the technical directing demands at the control room switcher during production.

*Designing the titling and graphics list for the production (P 18).* The producer designs the complete order and content for all titling and graphics for the production. This requires some aesthetic design because style and color are part of the choices available for character generator or computer designed graphics. (See the graphic design form and the character generator copy form.)

*Choosing and meeting with the assistant director and the production assistant (D 20)* The director continues to add to the production crew by choosing an assistant director and a production assistant. The director then meets with them. The director informs the assistant director of the responsibilities of the assistant director to the production and to the director. Most of that role will take place during production. The director conveys the director's expectations of the production assistant. The character generator slate, title, and other screen text are the responsibility of the production assistant.

*Meeting with the director (AD 1)* The assistant director meets with the director. The assistant director should have a sense of how the director sees the assistant director's role during production. A primary component to assisting the director is to help control and organize the blocking and spiking of the actor(s) and actress(es). This is accomplished by using colored, adhesive dots. These have to be purchased and a different color assigned to each actor and actress. Cameras also need to be assigned a color for spiking their positions.

*Meeting with the director (PA 1)* The production assistant meets with the director and receives the director's expectations of the production assistant for the production. The production assistant has important preproduction responsibilities. The production assistant is responsible for entering all of the copy into the character generator before production begins. The character generator copy includes the academy leader slates for every take, commercial titling, and screen texts.

*Choosing and meeting with the floor director and the continuity person (D 21)* The next production crew members chosen by the director are the floor director and the continuity person. The floor director becomes the director's head and hands in the studio during production. The continuity person is responsible for observing and recording the production details of the set(s), properties, dialogue, and actor(s)/actress(es) to reestablish them from the production of one script line unit to another.

*Meeting with the director (F 1)* The floor director meets with the director as the first step in preparing for the production. The expectations of the director are made clear during this meeting. In studio commercial production the floor director functions as a stage director, talent call person, line prompter, and referee among conflicting artists.

*Meeting with the director (CP 1)* The continuity person meets with the director to learn the expectations of the director with regard to production details being observed and recorded.

*Choosing and meeting with the video recorder operator (D 22)* The final production crew appointment is that of the videotape recorder operator. The director chooses and meets with the videotape recorder operator. The videotape recorder operator is responsible for securing videotape stock to videotape the entire commercial and supplying master tapes for postproduction editing. All these tapes should be striped with SMPTE time code for ease in editing in postproduction.

*Meeting with the director (VTRO 1)* The videotape recorder operator meets with the director and receives preproduction assignments. The videotape recorder operator is responsible for securing videotape stock for videotaping the commercial. Videotape stock should also be secured for postproduction editing of the master tape.

*Preparing the shot list breakdown from the master script (D 23)* The director prepares a shot list for each camera operator from the master script. The shot list is a listing of each planned camera shot in chronological order but broken down for each camera operator. The shot list contains the number of the shot for each camera, the shot's assigned framing, the content and composition of the shot, and any camera movement required.

*Securing colored, adhesive dots (AD 2)* The assistant director secures or purchases enough adhesive dots of different colors for each cast member and camera.

*Assigning colors to each actor, actress, and camera (AD 3)* The assistant director creates a master list of cast members on which to record the assigned color for each cast member and each studio camera to be used during blocking and videotaping. These spike marks have to be placed on the studio floor by the assistant director during blocking and removed after a studio wrap.

*Securing titling and screen text copy from the producer and entering the copy into the character generator (PA 2)* The production assistant secures the titling and screen text copy from the producer and begins to enter the copy into the character generator and records the information. The production assistant also enters and records the academy leader copy. (See the character generator copy form.)

*Preparing forms for recording continuity details during production (CP 2)* The continuity person prepares forms for recording sets, properties, dialogue, and actor/actress continuity details. (See the continuity notes form.)

*Determining and securing the videotape stock required for production and postproduction (VTRO 2)* The videotape recorder operator orders and purchases the videotape stock necessary to cover the production and postproduction editing needs.

***Checking studio cameras (VEG 1)***   The video engineer, while not always considered a member of the immediate production crew, plays an important role in videotape production. At some stage of preproduction, the video engineer checks the studio cameras for proper operation. This includes a physical check of the studio hardware as well as an electronic check from the video control area.

***Performing preventive maintenance on the studio cameras (VEG 2)***   The video engineer performs any scheduled preventive maintenance on the studio cameras. This maintenance includes checking the balance of the camera head on its pedestal, checking pan and tilt friction, and cleaning the camera lens.

***Checking the teleprompter monitors on the studio cameras (VEG 3)***   The video engineer also checks the functioning of the teleprompter as part of the check of the studio cameras.

***Reviewing the script for talent prompting (TO 1)***   Should the talent require script copy prompting, the prompter operator secures the prompter copy from the producer or director and reviews it in preparation for production.

***Striping SMPTE time code and coding and labeling all videotape stock (VTRO 3)***   The videotape recorder operator stripes all videotape stock with SMPTE time code to make postproduction editing easier and faster. After striping, all videotapes should be coded for ease in accounting for source tape(s) later and the videotapes and their cases accurately labeled.

***Checking the operation of the videotape recorders to be used during production (VTRO 4)***   The videotape recorder operator checks the operation of the videotape recorders to be used during production. The record video- tape deck and any playback deck (e.g., on which to B-roll stock videotape footage) need to be in the best condition possible so as not to waste production time.

***Performing preventive maintenance (VTRO 5)***   After checking the operation of the videotape recording decks, the videotape recorder operator performs any scheduled preventive maintenance on the decks. At a minimum, this includes cleaning the record heads.

***Completing set(s) construction and previewing the set(s) with the director (SD 7)***   The set designer over- sees the completion of the set(s) required for the produc- tion and gives the director a preview of the completed set(s).

***Previewing the completed set(s) with the set designer (D 24)***   The director previews the completed set(s) con- struction with the set designer and suggests changes or gives final approval.

***Lighting the studio set(s) (LD 5)***   Once studio set(s) are complete, the lighting director begins placing lighting

instruments, aiming them, and setting light intensity to create the lighting plot for each set.

***Reviewing the light pattern on the studio set(s) with the director (LD 6)***   The lighting director reviews the light pattern on the studio set(s) with the director when the lighting design is completed. This review is only preliminary because no true judgment of a lighting design can be made until the cast, in costume and make-up, take their places within the light pattern. Final judgment of the light pattern will be made before the studio videotape session.

***Previewing the preliminary lighting design on the set(s) (D 25)***   The director previews, makes any changes to, and approves the preliminary lighting designs(s) on the studio set(s). This approval is only of preliminary lighting because final lighting can be judged only with actor(s) and actress(es) in place and on-camera over control room monitors.

***Securing master videotapes for A-roll and B-roll requirements (VTRO 6)***   The videotape recorder oper- ator secures whatever videotape stock is necessary for the videotape session. The fresh stock was already striped and labeled. If B-roll videotape is going to be used (e.g., stock videotape, computer graphics, or animation), the videotape recorder operator has to secure the videotapes from the producer.

***Testing make-up, actor(s), and actress(es) under the studio lights on-camera (MA 6)***   The make-up artist tests all make-up design(s) on the actor(s) and actress(es) in the studio, under the lights, and on-camera. These tests can be videotaped for the director's later preview.

***Reviewing and demonstrating acquired properties to the director (PM 7)***   When all properties are collected, the properties master/mistress reviews and demonstrates the properties for the director. The director needs a working knowledge of the handling and operation of all properties. The properties operation will have to be approved by the director.

***Previewing properties (D 26)***   The director previews demonstrations of acquired properties. The director needs a working knowledge of how properties work or handle before blocking actor(s) or actress(es). The director makes final changes or suggestions and approves the properties for the properties master/mistress.

***Parading the final costumes before the director (CM 7)***   When costumes are completed, the costume master/mistress parades the costumes for the review of the director under studio lights and on-camera.

***Parading in costumes for the director's review and approval (T 5)***   The actor(s) and actress(es) must parade in costume for review and approval or change by the director.

***Reviewing costumes with the cast and costume mas- ter/mistress (D 27)***   The director reviews the costume parade under studio lights and on-camera with the cast and costume master/mistress. The director makes changes or suggestions and approves the costume.

*Reviewing make-up design(s) and tests for the director (MA 7)*   The make-up artist reviews make-up design(s) under studio lighting and in on-camera videotaped tests for the director.

*Reviewing the make-up design(s) with lighting and camera tests with the make-up artist (D 28)*   The director reviews the video tests of all make-up design(s) and lighting tests. The director makes suggestions or changes and then approves the tests.

## THE PRODUCTION PROCESS

For what appears to be a very brief unit of time—10, 15, 20, or 60 seconds—more cost, time, and attention go into the production of the television commercial than into the production of any other television genre. If the same cost, time, and attention were paid to the same time units of the other television genres (e.g., drama), the cost would become exorbitant and production prohibitive. There is no other television genre so orchestrated frame by frame, so calculated to effect audience response, or so precise in its use of time as the production of the television commercial. Consequently, for such a brief video unit, much studio production time is invested. An 8-hour production day may be scheduled for as little as a 60-second final product. And, still, the commercial may not be complete. The video produced in an 8-hour studio production session may have to be edited in postproduction. That editing may consume another 8-hour work session.

The commercial is still the heart and soul of television broadcasting and cablecasting. The production of the commercial appears, then, to merit the cost, time, and attention it receives.

## • Personnel

The production crew for the commercial does not differ from the role labels of any other type of studio production. Role definitions do differ, however. As in commercial preproduction, the engineering support staff will be considered among the studio production personnel.

*Producer (P)*   The producer's role during production is minimal, but supervisory. Working relationships occupy the producer more than anything else: cast and crew relations, and personnel and facility relations. The producer has the creative and aesthetic authority to accept or reject videotaped takes during production.

*Director (D)*   The director is at the hub of commercial production, just like the center of a wheel from which all spokes emanate. In charge and in control, the director directs the cast and crew from the studio before videotaping and from the control room during videotaping.

*Technical director (TD)*   The technical director works the switcher in the control room and responds to the take calls of the director during videotaping.

*Camera operators (CO)*   The camera operators and cameras are true players in the commercial being produced. Each camera and its operator has a defined position, spiked on the studio floor. Each camera has exact framing and shot composition requirements. The movement of the camera is assigned and coordinated with talent lines or blocking. In commercial production, the camera and operator are considered as cast members.

*Lighting director (LD)*   Another artist in commercial production is the lighting director. It is said that lighting makes television pictures possible. Lighting also creates mood, atmosphere, and the proper environment. The lighting director cannot simply light a set and sit back and watch the production. In the television commercial,

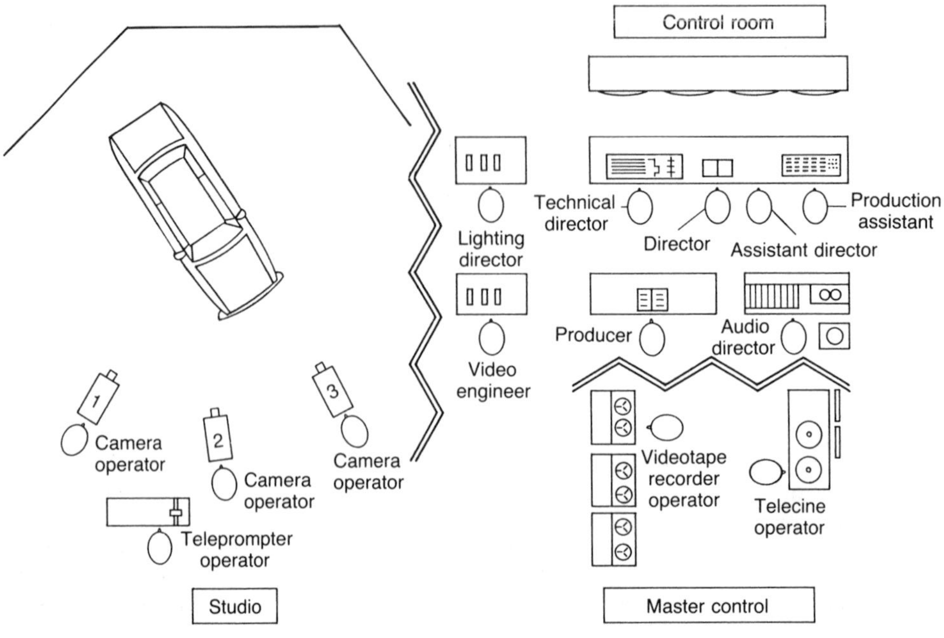

**FIGURE 2–19**
**Commercial production personnel placement.** This diagram shows each production personnel in place during production sessions.

shadows become a mar on the portrait. The lighting director is constantly vigilant during production to notice shadows and to control them.

*Audio director (A)* The audio director has a challenge in the studio production of a commercial. Although sound control is better in a studio, sound coverage is a challenge. Picking up talent without being seen either physically or by shadow is the test. Essential sound elements to any commercial are the artistic contribution of the audio director—music and sound effects.

*Assistant director (AD)* The assistant director's role, that of support, is most similar to the same role in any other studio production. The assistant director works closely with the director both in the studio (with the cast before videotaping) and in the control room (during videotaping with the master script).

*Production assistant (PA)* The production assistant performs generally as the role might be defined in any other studio production—that of maintaining character generator copy and memory recall. The added requirement is the constant need for the slate before every new videotape take. In addition, the slate requires constant update as videotaped takes progress and script line units change.

*Floor director (F)* The floor director assumes more responsibility in commercial production than does the equivalent role in other television genres. As the head and hands of the director in the studio during videotaping, the floor director will have the cast and support staff as well as the production crew to supervise. This often entails maintaining control between two sets of artists— the production and technical types and the creative and dramatic types. Once the director leaves the studio for the control room, the floor director will have to communicate more than production details to studio personnel: It often means interpreting the director's creative control to the cast and studio production crew also.

*Continuity person (CP)* The continuity person has the production responsibility to record all talent blocking, dialogue, costume, make-up, set, and properties details during videotaping. The continuity person then has the responsibility to reestablish those details for any subsequent videotaping of the same take or following script unit.

*Microphone boom grip(s)/operator(s) (MBG/O)* Additional crew, not generally found in other television genres, is the microphone boom grip or operator. With the extensive movement of actors and actresses in large sets, the task of sound coverage is compounded. Microphone boom operators are necessary if the studio uses a platform boom; microphone grips will be needed if the sound coverage will be accomplished by handheld extension booms or fishpoles. The compounding challenge to the microphone boom grips or operators is keeping themselves, their hardware, and their shadows from the camera lenses.

*Videotape recorder operator (VTRO)* The videotape recorder operator has the preproduction responsibility to ready all playback videotape decks for playback video sources and to assign a videotape record deck. Control room video and audio signals will have to be routed and checked. There will be constant need at the videotape recorder position for many possible short videotaped units required during commercial production.

*Video engineer (VE)* The video engineer is responsible for the proper maintenance of the studio cameras and teleprompter monitors on the front of the cameras. Uncapping cameras, routing the external signal to the cameras, checking tally light operation, and setting camera lenses to studio set light levels over cameras are all part of the video engineer's preproduction tasks. The more dynamic movement of multiple studio cameras under creative lighting design may be an added challenge for the video engineer during commercial production.

*Teleprompter operator (TO)* In some commercial productions, the talent will be expected to read the script copy from a teleprompter. In that case, a prompter operator will be required.

*Talent/actor/actress (T)* The talent in television studio commercial production has different requirements than in other genres. The talent meets at cast call without having been blocked (as is the custom in theater). In commercial production, the process of videotaping, not the actors and actresses, is the center of attention. In the studio, hardware (i.e., studio cameras) become players in the production. The big challenge to actors and actresses is to maintain character in the midst of what will appear to be technological confusion. Maintaining character will be further challenged by the constant take and retake of script line units. Another challenge will be the requirement to hurry up and wait. Being around but out of the way will test the patience of both cast and crew. Needing to be elsewhere to change or correct costuming, make-up, and properties, yet still be ready to be blocked or videotaped, will be difficult for the cast and crew. Picking up lines, blocking, and maintaining character in the middle of dialogue will be difficult for even the seasoned actor and actress.

*Costume master/mistress (CM)* The costume master/mistress has work that is well defined. Overseeing the costumes and accessories, dressing actors and actresses, cleaning and repairing costumes, tracking accessories, and checking costume detail are the responsibilities of the costume master/mistress.

*Properties master/mistress (PM)* The properties master/mistress has much to do during production. Supplying the set props, action props, and hand props is the responsibility of the properties master/mistress. The only time that props are not the responsibility of the properties master/mistress is while they are being used. Immediately before and after use, the props are in the care and security of the properties master/mistress. Maintaining an ordered storage area in the studio near the set(s), a place assigned for each prop, and a checklist of every property required is the responsibility of the properties master/mistress.

***Make-up artist (MA)*** The make-up artist has production responsibilities that do not end until the talent has left the studio after a production session. Generally, the talent does not get into make-up until after blocking and rehearsal—during the break between rehearsal and the first videotape take. Make-up that requires extensive application time (e.g., for aging or deformity) can begin before blocking. While most actors and actresses may apply their own make-up after testing in preproduction, the make-up artist is still needed for touch ups and details. In the studio, the make-up artist is constantly needed for make-up repairs for all actors and actresses.

***Set designer (SD)*** The set designer is not usually required to be in the studio during production. However, having designed the set(s), the set designer may choose to be in the studio at least for the first videotaping of each set. This gives the set designer the opportunity to dress the set according to the approved preproduction design(s) for the first time each set is used.

- **Production Stages**

***Supervising cast and crew calls (P 1)*** The producer begins the production process by supervising the cast and crew calls scheduled on the production schedule form. The crew and cast calls are scheduled gatherings or rendezvous times for production personnel. Supervisory personnel often use the crew call as the time for a final production meeting. The cast and crew calls may be a joint meeting time or separate. Some actors and actresses may need a lot of lead time in order to be ready (e.g., for detailed make-up preparation) for production, which may entail an earlier cast call. On the other hand, the production crew needs enough lead time to be at a point of readiness when the talent arrives. Studio production is often characterized as a "hurry up and wait" time. Adequate crew call lead time can remedy that for the cast. This is an opportunity for the producer and director to convey last minute details or changes in production that pertain to the crew and cast.

***Meeting cast and crew calls and holding a production meeting (D 1)*** The director begins production by meeting with both the cast and crew at either joint or separate calls. The director takes this opportunity to hold a production meeting in which the director sets expectations for the day's production, announces changes to the master script or to any other production element, announces crew changes or facility differences, and fields questions from the crew or cast.

The director concludes the production meeting by setting a time convenient to the crew for set-up and the cast for costume and make-up preparation for the first blocking of the day. Except for make-up artists required for extensive make-up application, only the production crew members must have their equipment ready for blocking.

***Meeting with the producer and director for crew call and production meeting (AD 1)*** The assistant director meets with both the producer and the director for crew call and production meeting.

***Handling studio facility and set(s) arrangement (P 2)*** The producer follows through with the role responsibility for the producer from preproduction and checks on expected arrangements (i.e., those arrangements determined and ordered during preproduction). This includes everything from air conditioning in the studio to set dressing. The producer initiated facility relations before production and must maintain the relationship with facility supervisory personnel during production. Any additional requests and changes regarding the facility should go through the producer.

***Supervising final studio facility and set(s) arrangements (D 2)*** The director walks through the studio production facility and the studio set(s) as a final check on details before and during the set-up for production. This affords the director the opportunity to make the producer aware of anything missing or not yet in place for production.

***Handling production crew and cast details (P 3)*** The producer handles personal details for the members of the cast and crew. This can entail anything from storage for valuables during protection to supplying hot water in the make-up room, coffee in the green room, or parking for personal cars.

***Supervising equipment set-up and placement (D 3)*** The director oversees all equipment set-up for the production. As soon as the director completes the review of crew positions and the checks of hardware and operational conditions, the director can begin blocking. The breadth of personnel and equipment that has to converge into the moment of beginning videotaping is so extensive that the director has to learn of missing, inoperative hardware, or difficulties, so the producer can trouble-shoot and solve the problems. The production crew needs to be encouraged to bring any difficulties in production elements to the attention of the director and producer as soon as possible.

***Meeting with the producer and director for cast call and the production meeting (T 1)*** The actors and actresses meet with the producer and director for the scheduled cast call and production meeting.

***Meeting with the producer and director for crew call and the production meeting (SD 1)*** The set designer meets with the producer, director, and crew for crew call and a production meeting. This is an opportunity for the set designer to give everyone some pointers on the expectations and use of the set(s) for the production.

***Meeting with the producer and director for cast call and the production meeting (PM 1)*** The properties master/mistress meets with the cast for cast call and production meeting with the producer and director.

***Meeting with the producer and director for cast call and the production meeting (CM 1)*** The costume master/mistress meets with the cast for the cast call and production meeting with the producer and director.

***Meeting with the producer and director for the cast call and the production meeting (MA 1)*** The make-up artist meets with the cast for the cast call and production meeting with the producer and director.

***Meeting with the producer and director for crew call and the production meeting (TO 1)*** The teleprompter operator meets with the producer and director if a script copy prompting device is needed for production.

***Overseeing the set(s) decoration and dressing (SD 2)*** The set designer, while not required during production, may choose to oversee the final touches on the set(s). This includes set decoration and dressing.

***Completing the set(s) decoration with set, hand, and action properties (PM 2)*** The properties master/mistress completes the set(s) decoration with the set, hand, and action properties needed for the day's production. Because most properties are not left lying around before and after production, properties have to be gathered from storage and security, laid out, and placed on the set before each production session. Many consumable props (e.g., food and beverages or cigarettes) have to be prepared or replenished.

***Meeting with the producer and director for crew call and the production meeting (CO 1)*** The camera operators meet for crew call and the production meeting with the producer and director.

***Meeting with the producer and director for crew call and the production meeting (TD 1)*** The technical director meets with the producer and director for crew call and the production meeting.

***Meeting with the producer and the director for crew call and the production meeting (A 1)*** The audio director meets with the producer and director for crew call and the production meeting.

***Meeting with the producer and director for crew call and the production meeting (MBG/O 1)*** The microphone boom grip(s)/operator(s) meet with the producer and director for crew call and the production meeting.

***Meeting with the producer and director for crew call and the production meeting (F 1)*** The floor director meets with the producer and director for crew call and the production meeting.

***Meeting with the producer and director for crew call and the production meeting (LD 1)*** The lighting director meets with the producer and director for crew call and the production meeting.

***Meeting with the producer and director for crew call and the production meeting (VTRO 1)*** The videotape recorder operator meets with the producer and director for crew call and the production meeting.

***Meeting with the producer and director for crew call and the production meeting (PA 1)*** The production

assistant meets with the producer and director for crew call and the production meeting.

***Meeting with the producer and director for crew call and the production meeting (CP 1)*** The continuity person meets with the producer and director for crew call and the production meeting.

***Lighting the set(s) and performing the lighting design check (LD 2)*** The lighting director lights the set(s) according to the approved lighting design. Given some passage of time from lighting design completion to the production session, the lighting director needs to check carefully that all prehung and preset lighting instruments are in place, aimed correctly, and set at the correct intensity. The lighting director may call the cast or crew to stand in for an actor or actress while lighting is checked.

***Uncapping the studio cameras (VEG 1)*** The video engineer begins production duties by uncapping the cameras. Cameras are uncapped physically (by removing the lens caps from the front of the camera's lens) and electronically (by opening the iris of the lens via video control).

***Readying the cameras for shading and videotaping (VEG 2)*** After uncapping the cameras, the video engineer supports the set by readying the cameras. The video engineer will help camera operators to prepare their cameras. The video engineer has engineering responsibility over the cameras, whereas the camera operators have operation responsibility.

***Setting up and preparing assigned cameras and placing them for first blocking (CO 2)*** The camera operators set up their respective cameras and prepare them for shading and first blocking by the video engineer. Preparation involves uncapping the lenses at the camera, adjusting pan and friction tension, setting or balancing the pedestal operation, adjusting control and zoom arms to a comfortable height, placing the camera in left to right (i.e., camera 1, camera 2, camera 3) order, and drawing sufficient camera cable behind each camera to allow for adequate movement. The video engineer requires that the cameras be aimed and focused on the same object in the lighted set for camera shading purposes.

***Setting up the teleprompter bed with the script copy (TO 2)*** If the studio script copy prompting device is needed for the talent during production, the teleprompter operator prepares the prompter bed with the script copy and tests it over the camera monitors.

***Preparing to run the script copy for talent reading during rehearsals and videotaping (TO 3)*** The teleprompter operator prepares the prompter bed and awaits the director's instruction to run the prompter during rehearsals and videotaping.

***Checking video levels of cameras with the lighting on the set(s) (VEG 3)*** The video engineer requires that the camera operators aim and focus their cameras on the same object within the lighted set. This allows the video

**FIGURE 2–20**
**Script copy prompter.** Many commercial productions require script copy prompting. Continuous script copy is fed under a light and a camera lens to be reproduced over a monitor on the front of the camera. The talent reads the script as reflected onto glass in front of the camera lens.

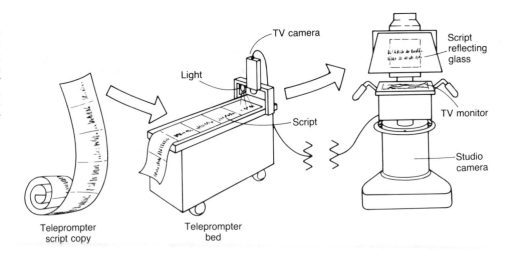

engineer to shade and set lens levels for optimum videotaping by the cameras. Shading requires setting contrast ratios between light and dark areas of the lighted set.

### Reviewing the shot list and attaching that list to the back of the camera (CO 3)

The camera operators review their respective shot lists prepared by the director and distributed during preproduction. The shot list has all of the prepared shots for each camera transcribed from the director's master script and listed by camera. While these lists are subject to change during blocking and rehearsal, they are a good point of reference for the camera operators. The more familiar with the demands to be made on the camera and the camera operator at this stage, the easier will be the blocking and rehearsal for the camera operators. The shot lists are attached behind the cameras and below the camera monitors.

### Routing the external video signal to the camera monitors for camera operator use during rehearsal and production (VEG 4)

It is an aid to camera operators to be able to see what the other camera operators are creating in their viewfinders. This information assists the setting of equivalent lens framing across cameras, especially in dialogue cutting between script lines. The video engineer routes the external video signal from the control room line monitor through the individual cameras, thus allowing the camera operators to choose to view the external video signal over their camera monitors.

### Alerting the camera operators and the director when ready (VEG 5)

The video engineer alerts the camera operators and the director when the cameras are shaded and ready for videotaping. The cameras may now be moved from their aimed and focused position.

### Meeting with microphone boom grip(s)/operator(s) and setting up the audio equipment (A 2)

The audio director meets with the microphone boom grip(s)/operator(s) and begins setting up the audio equipment.

### Meeting with the audio director and beginning audio equipment set-up (MBG/O 2)

After the production meeting, the microphone boom grip(s)/operator(s) meet

with the audio director to begin setting up the audio equipment.

### Assembling the microphone booms with microphones and running audio cables (MBG/O 3)

The microphone boom grip(s)/operator(s) begin assembling whatever studio sound coverage equipment remains to be assembled following preproduction. Microphone cables have to be run from the set(s) and connected to the wall or box connector.

### Running microphone cables and recording studio microphone inputs (A 3)

As microphone boom grip(s)/operator(s) complete cable runs, the audio director records the wall or box input connections in the studio for the microphones.

### Patching studio microphone inputs into the audio control board in the control room (A 4)

The audio director patches the corresponding input selection for microphones from the studio into the audio board in the control room.

### Making intercom network connections (A 5)

The audio director selects those intercom network connections needed during production. The audio director will need principal communication with the director.

### Preparing and assisting the director with the master script (AD 2)

The assistant director assists the director, primarily with the master script. The assistant director must have the director's copy of the master script and must follow the director closely during preblocking stages of production.

### Assuming responsibility for the studio, the set(s), the cast, and the crew during studio use (F 2)

As studio set-ups progress, the cast prepares, and other principals (e.g., set designer, properties master/mistress) relinquish their primary obligations, the floor director takes control and responsibility within the studio. This control begins with awareness of the state of preparation of all elements of the studio. The floor director wants to be able to report studio readiness to the director when asked.

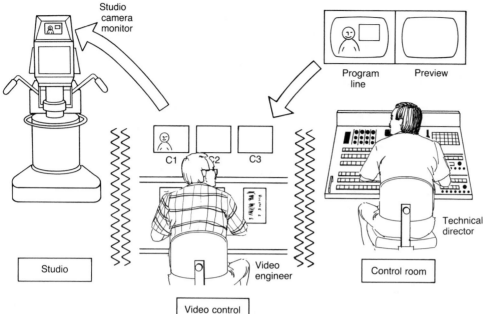

**FIGURE 2–21**
**External video signal.** The external video signal is routed from the control room program line back to the camera monitors. This permits each camera operator to see what the other cameras are framing and to adjust their framing to match.

***Checking intercom network connections (F 3)*** The floor director checks the intercom connections from the director and the studio personnel.

***Checking intercom network connections (TD 2)*** The technical director checks the intercom network connections from the videotape recorder operator, the director, and the production assistant.

***Checking intercom network connections (PA 2)*** The production assistant checks the intercom network connections needed at the character generator. The production assistant needs contact with the technical director, assistant director, and director.

***Knowing what the director needs and requires during rehearsals and takes (F 4)*** It is important to the production process and to the role of the floor director to know at all times what the director needs and requires during all stages of production. This includes knowing where the director intends to begin with the script, any changes in production order, and the cast required for any stage. Much of this information is contained on the production schedule.

***Entering and recording character generator copy for script units to be produced (PA 3)*** The production assistant enters any character generator copy that was not entered and recorded during preproduction. The production assistant prepares for the last minute needs of the character generator for script line units about to be produced. This preparation includes not only the updated slate for the academy leader for videotaping but also any titling or screen text to be put onto videotape during any script line unit to be produced. This information is found on the master script and on the production schedule that was distributed during preproduction and updated during the production meeting.

***Preparing the record videotape deck and selecting labeled videotape stock (VTRO 2)*** The videotape recorder operator prepares the record videotape deck for recording production takes by selecting the appropriate videotape stock assigned for the production. Any B-roll needed is also prepared during this stage of production. Exterior stock video footage, for example, may be required by the script to be inserted into the video.

***Preparing for A-roll and B-roll video (VTRO 3)*** The videotape recorder operator readies for the start of production by preparing videotapes and videotape decks for both A-roll (the primary record videotape deck) and B-roll (the secondary or inset video-to-video playback deck). This will be necessary only if prerecorded video (e.g., stock shots) will be inserted into the commercial during production.

***Choosing videotape playback and record videotape decks and alerting the technical director of the signal routing assignment through the production switcher (VTRO 4)*** Once the videotape recorder operator knows the videotape requirements for the production, the assignment of videotape decks for both record and playback have to be made. After videotape decks are assigned, the videotape recorder operator alerts the technical director of the assignment of the video signal through the switcher in the control room.

***Noting the assignment of the video signal from the playback videotape deck through the production switcher (TD 3)*** The technical director notes the assignment of the video signal from the videotape recorder operator for the playback B-roll videotape deck through the production switcher.

***Readying forms for recording continuity notes during rehearsals and takes (CP 2)*** The continuity person

**FIGURE 2–22**
**Routing A-roll and B-roll video sources.** The videotape recorder operator assigns any video source inserts to A-roll and B-roll videotape decks. These signals are then routed to the control room for the technical director's control through the production switcher.

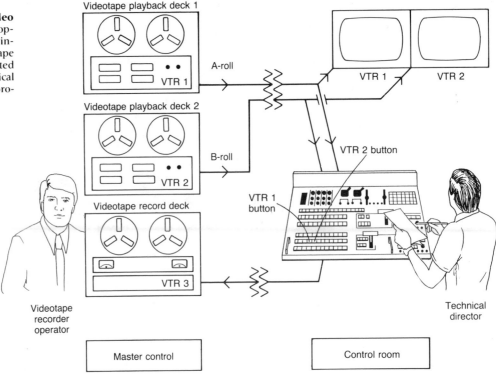

readies continuity recording forms for the script line units scheduled for the production session. Readying the continuity forms requires labeling individual forms for the specific script line units to be blocked, rehearsed, and videotaped. Relevant information for the continuity forms is available from the production schedule form and from the videotape recorder operator. The videotape recorder operator has the code(s) for the videotape stock and source tape(s) being used for the script line units recorded during the production session.

***Meeting the lighting director's requirement for lighting stand-in needs (T 2)***  The actor(s) and actress(es) meet the lighting director's requirements for standing in for final lighting instrument aiming and intensity setting during preblocking preparation of the set(s). There comes a point in lighting when the person to be lighted is required as a stand-in for final adjustments (to adjust for such attributes as height and coloring).

***Checking intercom connections (VTRO 5)***  The videotape recorder operator checks the required intercom connections for master control. The videotape recorder operator must be in contact with the technical director, assistant director, and director.

***Checking intercom network connections and announcing readiness (CO 4)***  The camera operators check that they have the correct intercom network connections (i.e., to the director and the floor director) and alert the floor director that they are ready.

***Checking intercom network connections for the director and the assistant director (AD 3)***  The assistant director checks any intercom network connections for the

director. The director requires intercom connection to the technical director, videotape recorder operator, production assistant, camera operators, floor director, audio director, and assistant director. The assistant director needs the same connections.

***Alerting the director when ready (VTRO 6)***  The videotape recorder operator alerts the director (or the assistant director) when the videotape recording equipment is ready and the videotape stock is threaded and cued to begin.

***Calling and readying actor(s), actress(es), and cameras by script line unit and spiking actor(s), actress(es), and cameras (AD 4)***  The assistant director calls and readies the cast and crew to the beginning of blocking. This call is a specific reminder about the exact script line unit being blocked. The assistant director readies the adhesive, colored dots for spiking the actor(s), actress(es), and cameras during blocking. The adhesive dots are color coded for each actor, actress, and camera.

***Following the director's instructions for blocking; rehearsal(s); videotape take(s); and calls for action, cut, or freeze (T 3)***  The cast should be prepared to follow without comment any directions that the director may give. Directions include the command to begin the action of the script line unit, to cut or stop at any point in a unit, or to freeze at any time. The freeze is called to enable the continuity person to record details of costuming, movement, or property use to facilitate editing any subsequent takes.

***Following the master script with and for the director (AD 5)***  The assistant director handles the master script and follows it for the director as the assistant director follows the director during blocking. Any changes the director spontaneously makes during blocking and rehearsal are recorded in the director's master script.

***Beginning blocking and blocking actor(s) and actress(es) without lines (D 4)***   When the appointed time for beginning blocking arrives, the cast and studio personnel should be ready. The assistant director accompanies the director in the studio on the set scheduled for the first script line unit. Camera operators, audio director, and microphone boom grip(s)/operator(s) should be ready to observe blocking. The director begins blocking.

Blocking is the first interface between the cast and crew for the purpose of videotaping script line units. The cast and crew are aware from the production schedule form of the script line units, set, cast, properties, and costumes required and of the studio blocking starting time. At this point, the cast is usually not in costume or make-up. Most commercial production processes take a break between the last rehearsal after blocking and the first videotaping for make-up and costuming. If the length of time needed for applying make-up or dressing in costume is excessive, make-up and costuming may have been begun and even been completed before blocking. (This is an exception.)

The director blocks actor(s) and actress(es) by physically moving each, one at a time, from a starting position within the set through a movement and to a resting position, line by line, stage direction by stage direction, as preplanned from the director's master script. At each resting position, the assistant director places an adhesive dot between the feet of each of the actor(s) and actress(es) as they are blocked. The dots are color coded for each cast member.

If the talent requires script copy prompting during rehearsal and production, the director makes use of the teleprompter operator and the studio prompting device.

***Rehearsing the blocking of the actor(s) and actress(es) without their lines (D 5)***   The director rehearses the actor(s) and actress(es) through the blocking *without* their lines. The director watches each cast member hit the assigned spike marks (colored dots).

***Rehearsing the actor(s) and actress(es) with their lines (D 6)***   The director now rehearses the actor(s) and actress(es) *with* their lines. The director watches and affirms or corrects the rehearsal. The director may repeat this step.

***Standing by to handle action properties before and after all rehearsals and takes (PM 3)***   The properties master/mistress stands by during all rehearsals and takes to handle and store all action props used by the actor(s) and actress(es). Many props will have to be replenished or reloaded, depending on the nature of the property. This is the responsibility of the properties master/mistress.

***Observing the blocking and rehearsal of the actor(s) and actress(es) by the director (CO 5)***   The camera operators observe the blocking and rehearsal details of the actor(s) and actress(es). This orients the camera operators to what will be expected from them during videotaping.

***Placing microphone boom grip(s)/operator(s) and microphone boom(s) for the script line unit being blocked (A 6)***   The audio director watches the blocking together with the microphone boom grip(s)/operator(s) for plan-

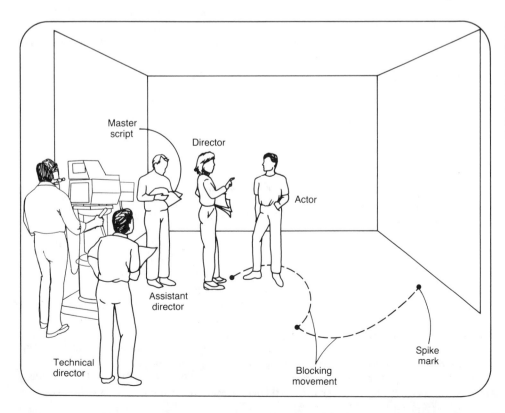

**FIGURE 2–23**
**Production blocking: Step 1.** The first step in blocking for videotaping is for the director to physically move the actor(s) and actress(es) from point to point. The assistant director spikes each rest point with a colored adhesive dot. During this step, the actor(s) and actress(es) do not deliver their lines.

Master script

Director

Actor

Assistant director

Technical director

Blocking movement

Spike mark

**FIGURE 2-24**
**Production blocking: Step 2.** The director rehearses the actor(s) and actress(es) first without lines, just hitting their spike marks, then while they deliver their lines on cue.

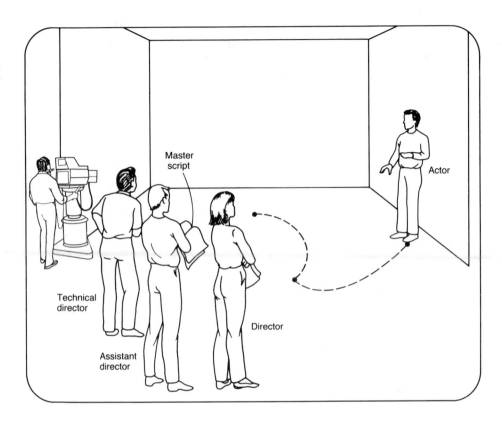

ning the placement of grip(s) and boom(s) to cover studio sound during rehearsals and videotaping.

***Taking place(s) for preliminary dialogue and sound coverage (MBG/O 4)*** The microphone boom grip(s)/operator(s) take their planned places in the set for preliminary dialogue and sound coverage. This step for the audio director is just a beginning attempt to find and test the best place for sound coverage.

***Blocking the cameras as actor(s) and actress(es) again hit their spike marks (D 7)*** The director blocks each camera physically as the actor(s) and actress(es) hit their spike marks again, usually without their lines, but sometimes by delivering their lines slowly.

The director takes each camera to its defined placement and has the assistant director spike the camera's placement with the coded adhesive dots. The spike marks for the cameras are usually placed directly under the center of the camera's pedestal trunk. The director then sets the defined composition (actor or actress) and its framing (e.g., CU or LS) for each shot. Any secondary movement (e.g., zooming or dollying) is blocked. All of these directions should correspond to the shot lists that the camera operator received from the director. The camera operators update their own shot lists, making any additional notes to assist them in getting each shot during rehearsals and takes.

***Being blocked with the actor(s) and actress(es) and spiking the cameras (CO 6)*** The cameras are now blocked. The camera operators should see their cameras and themselves as other actors in the script line unit. They have to be in their defined places with the assigned

camera composition and framing just as the actor(s) and actress(es) are required to be on their defined spike marks when dialogue lines are delivered and movements are made. The assistant director spikes each camera placement and marks it with a colored adhesive dot.

***Noting details of camera placement and shot composition and framing and making changes on the shot list (CO 7)*** Each camera operator makes notes on the shot list for the camera and includes any details of the placement of the camera, the shot composition, and the assigned framing. Any changes and additions to the shot list should be made as each shot is assigned by the director. Each camera operator should also make notes for each shot to help in getting the required shots again during rehearsals and takes.

***Rehearsing cameras, actor(s), and actress(es) (D 8)*** The director rehearses the cameras and cast for the first time by loudly calling out in advance the number of each defined shot. (The assistant director may be asked to call the shot numbers out for the director.) This may be the most tedious step in the production process as the cast and crew attempt to integrate and coordinate their blocking. The cast may be asked to slow the pace of delivery and movement for this step.

As this rehearsal progresses, the audio director and the microphone boom grip(s)/operator(s) also attempt to integrate their sound coverage.

The lighting director begins to monitor lighting from the shots viewed over the cameras.

The director may move from camera monitor to camera monitor to check the defined shots. The director relies on

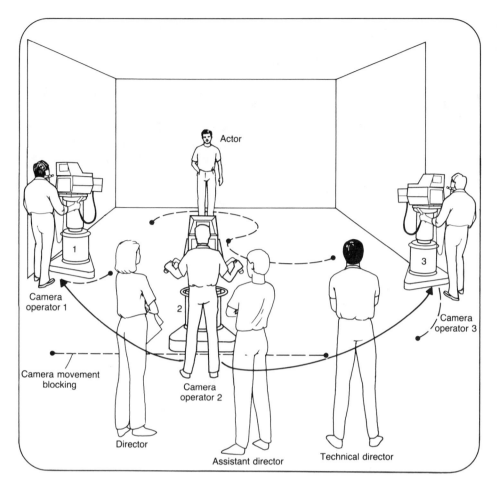

**FIGURE 2–25**
**Production blocking: Step 3.** The director blocks each camera by moving each to its mark, spiking it, and setting the framing of the lens and its content. Actor(s) and actress(es) just hit their spike marks for this step unless otherwise directed.

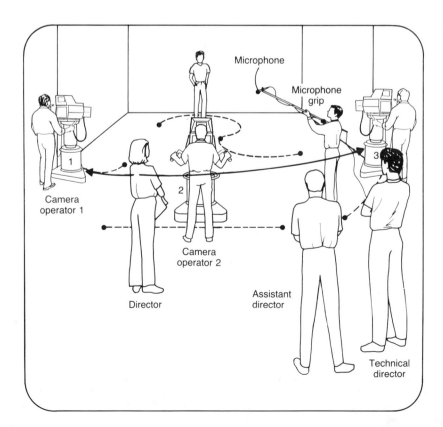

**FIGURE 2–26**
**Production blocking: Step 4.** The director rehearses the cameras and actor(s) and actress(es). At this stage, microphone coverage can be added. The director moves from camera to camera as shots are called loudly by the assistant director. This step may be repeated several times.

feedback from each camera operator regarding individual defined shots and the operator's ability to get the shots as designed.

***Rehearsing cameras with actor(s) and actress(es) (CO 8)*** The camera operators rehearse their defined shots as the actor(s) and actress(es) rehearse their dialogue and movement. The camera operators follow their shot lists scrupulously. This is the time to discover any difficulty in getting assigned shots, in moving the cameras, and in framing and focusing.

***Making changes in blocking and continuing rehearsal in the studio (D 9)*** The director listens to the camera operators and cast on the difficulties they may be having in achieving their defined blocking. Then, either by trouble-shooting the difficulties and retaining original blocking or by changing the blocking, the director rehearses from the studio again. The director continues these steps until the cast and crew achieve what the director requires.

***Updating the shot list (CO 9)*** The camera operators update and make any changes to their camera blocking, in the numbering of shots, or in the actor(s) or actress(es) on their shot lists. Camera operators should inform the director of all difficulties they experience in getting the assigned shots and make changes as the director instructs them.

***Rehearsing as often as the director calls for rehearsals (CO 10)*** The camera operators rehearse as often as the director calls for rehearsals, getting all assigned shots and details of shots as directed.

***Responding from the control room switcher to the director's calls for shots from the studio (TD 4)*** The technical director is in the control room at the switcher. The technical director responds to the director's calls from the studio and cuts the camera shots as called by number. This allows the director to watch the studio floor monitor for the succession of camera shots as designed.

***Watching the camera and cast blocking for microphone sound coverage and conferring with the audio director (MBG/O 5)*** The microphone boom grip(s)/operator(s) watch camera and cast blocking and rehearsals for microphone sound coverage possibilities. The grip(s)/operator(s) confer with the audio director for best microphone placement and try sound coverage during rehearsals.

***Beginning microphone sound coverage during camera blocking with actor(s) and actress(es) lines (MBG/O 6)*** The microphone boom grip(s)/operator(s)

**FIGURE 2–27**
**Production blocking: Step 5.** At this step, the technical director retires to the control room to cut from shot to shot as the rehearsal progresses. The shots are called from the studio as the director watches from the floor monitor. Changes and corrections are made and rehearsed as often as the director deems necessary. The director and assistant director then retire to the control room for a videotape take.

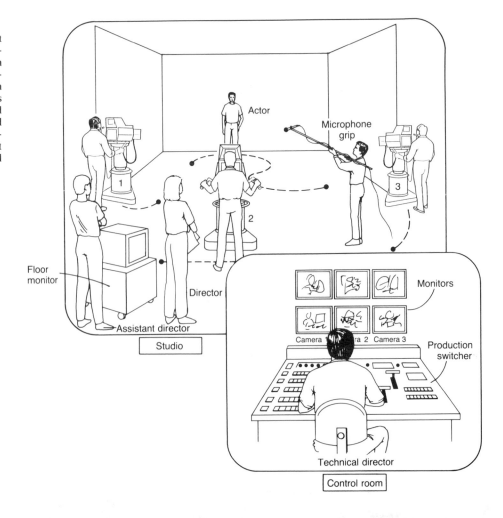

begin sound coverage as soon as the director rehearses the actor(s) and actress(es) with lines. The best thing the grip(s)/operator(s) can do is attempt sound coverage. Given the many factors and difficulties of sound coverage, the sooner they attempt to cover dialogue, the sooner they can work out problems. Besides sound coverage, the grip(s)/operator(s) have to avoid getting on-camera, catching their microphones on-camera, or casting shadows on the set. Problems of coverage have to be identified and solved as soon as possible.

***Conferring with the audio director for improved microphone placement and sound coverage and making changes to avoid getting on-camera or casting shadows (MBG/O 7)*** The microphone boom grip(s)/operator(s) confer continuously with the audio director for improved microphone placement and sound coverage during all rehearsals. The grip(s)/operator(s) make those changes that get them and their equipment off-camera and eliminate any shadows.

***Noting CU/LS camera framing for improved sound perspective placement of microphone(s) (MBG/O 8)*** The microphone boom grip(s)/operator(s) note when cameras have close-up or long shot framing assigned to them. This allows the sound coverage to begin to achieve better sound perspective. When a shot is a close-up, the audience should be able to perceive dialogue sound closer. This can be achieved by placing the microphone closer to the actor or actress. Because the shot is framed tighter, the microphone can safely be placed closer without its being seen on-camera.

If a camera shot of an actor or actress is longer (i.e., a long shot), the audience expects more distant sounding dialogue. Because the shot is framed longer, the microphone has to be placed farther from the actor or actress to avoid being caught in the camera lens. This should give better sound perspective to the audience. Good grip(s)/operator(s) begin to memorize dialogue and know camera shots during rehearsals as part of their better sound coverage techniques.

***Making microphone audio level checks as needed (A 7)*** The audio director makes microphone sound level checks (i.e., volume, equalization, and tone) during rehearsals as often as possible to achieve acceptable levels. Difficulties in achieving levels will require trouble-shooting from the studio with the microphone boom grip(s)/operator(s).

***Making studio foldback sound checks for the microphone boom grip(s)/operator(s) and studio cast (A 8)*** The audio director patches and checks foldback sound for the microphone boom grip(s)/operator(s). The studio grip(s)/operator(s) monitor the dialogue from the cast over headsets during sound coverage throughout the production session.

***Monitoring cast dialogue with foldback over headsets during rehearsals and takes (MBG/O 9)*** Microphone boom grip(s)/operator(s) monitor their own sound coverage pickup by audio foldback over headsets. This

permits them a better sense of their own sound coverage and microphone placement during rehearsals and takes.

***Rehearsing prerecorded audio tracks and sound effects into the studio (A 9)*** The audio director interfaces any required sound effects (e.g., telephone ringing or doorbells) into the rehearsals whenever the director calls for them to be integrated. The cast has to hear prerecorded audio tracks during production. Speakers in the studio have to be checked for sound reproduction quality.

***Watching for microphone(s), microphone boom(s), and shadows on shots during rehearsals (CO 11)*** Camera operators watch their camera monitors closely for signs of any microphone(s), microphone boom(s), grip(s), or their shadows in any shot. The audio director is alerted if any of these appears in the shots.

***Observing camera and cast blocking and directing microphone boom and grip placement for better sound coverage (A 10)*** The audio director constantly observes camera and cast blocking and directs microphone boom and grip placement for better sound coverage of dialogue. The audio director observes camera and cast blocking either from the audio control board monitors or from the studio.

***Watching the set(s), actor(s), and actress(es) for light and shadows (LD 3)*** The lighting director watches from the studio or from the control room monitors for any problems in lighting the set(s) or the cast during rehearsals. The lighting director can anticipate lighting needs to a point. Once blocking occurs and rehearsals begin, problems of overlighting, underlighting, or excessive reflection and glare can appear.

***Adjusting lighting instruments or lighting intensity after each rehearsal (LD 4)*** The lighting director makes any changes in lighting instruments or lighting intensity between rehearsals.

***Rehearsing with actor(s), actress(es), and cameras (MBG/O 10)*** The microphone boom grip(s)/operator(s) begin rehearsing their microphone placement and sound coverage as soon as and as often as the cast and cameras rehearse. It is always more convenient to integrate microphone sound coverage as others rehearse than needing a separate rehearsal for sound coverage alone. Many rehearsals may be needed to trouble-shoot the problems associated with sound coverage.

***Making necessary changes and adapting (MBG/O 11)*** The microphone boom grip(s)/operator(s) make the necessary changes in sound coverage as problems are solved and adapt sound coverage placement and techniques when necessary. Grip(s)/operator(s) should consider all possible microphone placement necessary to cover sound (e.g., over the top of a set or under the camera's lens).

***Monitoring microphone placement and sound levels during rehearsals (A 11)*** The audio director is responsible for the microphone placement and monitors where microphone boom grip(s)/operator(s) are during rehearsals. The audio director constantly monitors and checks sound levels during rehearsals.

*Making microphone grip(s)/operator(s) placement changes before subsequent rehearsals (A 12)* The audio director has the final authority for the placement of grip(s)/operator(s) and microphone(s) during rehearsals and in preparation for videotaping.

*Mixing in any prerecorded sound tracks or music effects as required by the script (A 13)* The audio director mixes in any prerecorded sound tracks, sound effects, or music as required by the script. Audio levels have to be preset for these audio sources.

*Monitoring cast dialogue and sound effects during rehearsals (A 14)* The audio director monitors all audio sources during rehearsals, including the cast dialogue and prerecorded sound effects during rehearsals.

*Retiring to the control room for a rehearsal (D 10)* When the director feels that there have been sufficient rehearsals and all necessary changes integrated, the director retires to the control room with the assistant director. The director proceeds with a rehearsal from the control room.

*Readying shots to the camera operators and taking shots to the technical director (D 11)* The director follows the master script as assisted by the assistant director. The director then readies each defined and rehearsed camera shot and calls the defined camera shots when they are to be taken by the technical director.

*Studying lighting effects over the studio floor monitors and making final changes (LD 5)* The lighting director takes the opportunity afforded by rehearsals to study the lighting design effects over the floor monitors as actor(s), actress(es), and cameras move about in the set. Any undesirable effects are changed between rehearsals.

*Making changes and updating the master script for the director (AD 6)* The assistant director updates the master script with any changes in cameras, shot numbering, or cast blocking. Any changes that the director may want to make at this stage are made from the control room over the intercom network.

*Calling a break for cast costuming and make-up (D 12)* When the director feels all blocking, lines, camera shots, and sound coverage are adequate, the director calls a break for the actor(s) and actress(es) to get into costume and make-up for videotaping.

*Meeting with the actor(s) and actress(es) for last minute instructions and changes in make-up design (MA 2)* The make-up artist meets with the cast immediately into the break for costuming and make-up to convey any last minute instructions on applying make-up and any changes in make-up design.

*Meeting with the make-up artist for make-up preparation and beginning make-up application (T 4)* The cast meets with the make-up artist and begins make-up application when the director breaks from rehearsal(s) for make-up and costuming time. Make-up application that is detailed and a slow and tedious process needs to be begun as soon as possible after cast call instead of waiting until the rehearsal break.

*Meeting with the actor(s) and actress(es) to distribute costumes (CM 2)* The costume master/mistress meets with the cast to distribute the costumes required for the script production unit.

*Meeting with the costume master/mistress for costume distribution (T 5)* After make-up is complete, the cast meets with the costume master/mistress for the distribution of costumes.

*Beginning costuming (T 6)* The cast begins costuming after applying make-up. Actor(s) and actress(es) usually costume after make-up. Occasionally (e.g., when close fitting costumes must be pulled over the head and face), actor(s) or actress(es) will have to costume themselves and then apply make-up. Costumes need to be protected from make-up stains. Tissues stuck into the neck of costumes prevent make-up from staining costume materials until the last possible minute before videotaping.

*Assisting actor(s) and actress(es) into costumes (CM 3)* The costume master/mistress is responsible for the proper dressing of the cast and for the correct accessories for each costume. The costume master/mistress helps or supervises the cast as they dress.

*Proceeding with make-up application (MA 3)* The make-up artist helps or supervises the application of make-up by the cast. Most actor(s) and actress(es) apply their own make-up, but need a supervisory eye for amount of make-up and correctness of design.

*Checking each actor's or actress's make-up before videotaping (MA 4)* The make-up artist checks each member of the cast for final make-up design before returning to the studio. The make-up artist is still responsible for the make-up designs approved in preproduction.

*Watching takes for additional make-up needs, changes, or repairs (MA 5)* The make-up artist monitors the videotaped takes for any make-up needs, changes, or basic powdering repair the cast may require between takes.

*Being available yet out of the way in the studio during subsequent rehearsals and takes (T 7)* Once costuming and make-up are complete, the cast returns to the studio and awaits the director and final crew preparations. This usually means some waiting. A perennial problem in production is that of the cast and crew getting in each other's way as final touches are made before a take. The cast should be readily available for any needs the director or the crew may have of them, but must also stay out of the way of crew task completion.

*Calling for a take or a final rehearsal (D 13)* When the cast is finished with costuming and make-up and has returned to the studio, the director calls for a final rehearsal or a videotaped take. A director may videotape the rehearsal without alerting the studio. A final rehearsal after costuming and make-up serves to check out the effects of the lighting on both. If the rehearsal or take is to be recorded, the director readies the videotape recorder operator, awaits the response that the recorder is ready, and calls the videotape to roll and then takes the slate for the unit.

***Cutting from shot to shot as the director calls for each take during videotaping (TD 5)*** The technical director cuts from shot to shot as the director calls for each take during the videotaped takes.

***Observing studio monitors for additional make-up needs, changes, and repairs (MA 6)*** The make-up artist observes the studio monitors for any additional make-up needs, changes, or repairs to the actor(s) and actress(es) under the lights. Changes and repairs are made between rehearsals and takes.

***Monitoring lighting of set(s), actor(s), and actress(es) over the control room monitors during take(s) (LD 6)*** The lighting director monitors lighting levels and cast lighting over the control room monitors during takes.

***Readying, rolling, and recording the videotape deck with the director's call (VTRO 7)*** The videotape recorder operator responds to the director's call to ready the record videotape deck. The videotape recorder operator then begins to roll the recorder deck on call, and confirms to the director that the deck is recording.

***Monitoring the video levels of the cameras during videotaping (VEG 6)*** The video engineer constantly monitors the video levels of the light entering the lenses of the cameras during videotaping. The motion of the cast and cameras during production will frequently occasion a burst of light into the lenses. For example, an actor turning a highly reflective property may create a strong light reflection. The video engineer has to manually close the iris of the lens to control the burst of unexpected light.

***Monitoring recording and playback videotape machines during videotaping (VTRO 8)*** The videotape recorder operator monitors constantly the videotape recorder and videotape playback machine (should any B-

rolling be required) for proper operation and monitors the video and audio input levels throughout the videotaping session.

***Preparing and changing relevant character generator slate information for all videotaped takes of any script line unit (PA 4)*** The production assistant prepares and readies the character generator for the slate copy or any other information needed by the director during a videotaping session.

***Shooting the camera for take(s) (CO 12)*** The camera operators respond to the director's call and shoot their shot lists for any takes required. Camera operators follow their shot lists carefully and note any changes or additions to their camera placement, shot composition, or framing. Camera operators are reminded that with the extensive camera movement that may be required, constant attention is required to maintain depth of focus with the camera lenses. Any change in distance between a camera (having moved) or an actor or actress (with blocking movement) requires rack focusing of the lens by the camera operator or at least some fine focusing control by the camera operator.

***Alerting the director to soft focused cameras during videotaping (VEG 7)*** The video engineer alerts the director and technical director that a camera may be soft focused. The video engineer has a closer watch on control room monitors of the cameras and can spot cameras with less than sharp focus. Alerting the director and technical director permits them to let the camera operator know in order to correct the focus.

***Recording any dialogue and mixing effects during takes while monitoring for any extraneous studio sounds during the takes (A 15)*** During any takes, the audio director records dialogue and mixes any required effects. The audio director also monitors the studio for any extraneous and unwanted sounds during videotaping.

***Covering actor(s) and actress(es) for sound during takes (MBG/O 12)*** The microphone boom grip(s)/operator(s) cover all actor(s) and actress(es) for sound during takes.

***Following and keeping track of the master script for the director at the directing console in the control room during rehearsals and takes (AD 7)*** The assistant director follows the master script for the director with a finger or pencil so that the director can find the spot in the master script where the actor(s) and actress(es) are at any moment. The director depends on the assistant director to keep track of the dialogue.

***Finishing a take and stopping videotape recording (D 14)*** At the end of a take, the director calls for a cut to the studio, a freeze, and a stop to the videotape recorder.

***Stopping the videotape recorder at the director's call (VTRO 9)*** The videotape recorder operator stops down the videotape recorder when the director calls for a videotaping stop.

Stand-by in the studio
Ready to roll tape
Roll tape
Ready bars and tone
Take bars and tone (:30)
Ready slate
Take slate (:20)
Ready black
Take black (:10)
Ready to come up on camera 2, audio
Countdown to studio: 5–4–3–2–1

**FIGURE 2–28**
**Directorial language.** The beginning of a videotaped take requires a series of specialized steps and the language to call for them. This is especially true if the commercial is to be videotaped in one take.

***Recording actor(s), actress(es), set(s), dialogue, properties, and costume details from unit to unit and take to take for reestablishing for a subsequent unit or take (CP 3)***    When the director calls for a cut and stops videotaping a take and calls for a freeze, the continuity person enters the set while the cast freezes. The continuity person is responsible for recording all details that may affect subsequent units, takes, and postproduction editing. Detail of actor(s), actress(es), set(s), dialogue, properties, and costumes have to be reestablished and matched for subsequent unit(s) and takes. The content of some edits will need to be matched during postproduction editing and the continuity details will have to be cut together.

***Releasing actor(s) and actress(es) after recording continuity details following a take or a wrap (CP 4)***    The continuity person releases actor(s) and actress(es) after recording continuity details. The cast is free to break only after the continuity person releases them.

***Double-checking the director's calls for rehearsal(s), retake(s), next unit for blocking and videotaping, studio wrap(s), and studio strike (AD 8)***    The assistant director checks and relays the director's commands throughout rehearsal(s) and videotaping. The assistant director knows what the director wants at any moment or what the director intends to do at all times. This is especially true for announcing subsequent script line units for blocking. The assistant director informs others and is the resource person for that information.

***Assisting the director in the control room and studio, keeping track of the production process, and updating the control room log of videotaped takes (AD 9)***    The assistant director assists the director as necessary in the control room and the studio. The assistant director must keep track of the production process for the director. The assistant director keeps the control room log of videotaped takes, which is a record of the source tape for postproduction editing needs. (See the videotape log form.)

***Effecting what the director requires while the director is in the control room (F 5)***    The floor director continues to function as the director's other self in the studio. This means that anything the director needs for or from the studio should be carried out by the floor director. Once the director leaves the studio for the control room, the director should not have to return except for some serious problem.

**FIGURE 2–29**
**Continuity note taking.** The continuity person has to be attentive to all details of a videotaped unit. These details include the actor(s), actress(es), dialogue, costuming, make-up, properties, and blocking.

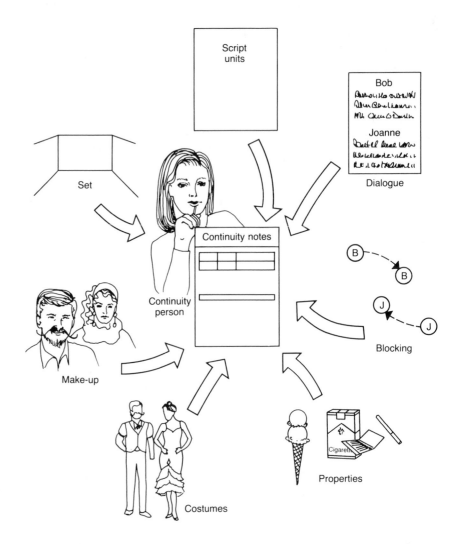

***Being informed and informing the cast and crew of the director's intentions (F 6)*** The floor director is informed by the director or assistant director over the intercom of the director's intentions at all times. The director's clear intentions can facilitate both cast and crew. The director may call another rehearsal, a take, a retake, a wrap; announce the next unit for blocking; or call a studio strike.

***Assisting the director's judgment or call for a retake or wrap (TD 6)*** At the end of a take, the technical director assists the director in judging the quality of a take by sharing insight and evident mistakes before the director calls a retake or wrap.

***Conferring with the producer on acceptable take(s) (D 15)*** The director confers with the producer on the acceptability of any take before announcing the next call. The decision to wrap a script line unit should be a mutual decision of the producer and director.

***Conferring with the director on the acceptability of a take (P 4)*** The producer confers with the director after each take and helps decide the acceptability of a take. The decision to wrap a script line unit should be a mutual decision of the producer and director.

***Replacing or replenishing properties consumed or used during rehearsal(s) or take(s) (PM 4)*** The properties master/mistress has to replace or replenish any properties that may have been consumed or used during a rehearsal or take. There are properties that will be eaten or smoked that will have to be replenished or relit before a subsequent rehearsal or take.

***Replacing or altering properties to the changing needs of the commercial during production (PM 5)*** During production, some properties will have to be changed or altered depending on the changing needs of the commercial and the script. Some consumable properties (such as ice cream cones and cigarettes) will have to be adapted to a later stage in consumption to match a previous unit. Cigarettes, for example, will have to be lit at a half smoked stage for a later unit.

***Watching rehearsal(s) and take(s) for the use, abuse, and needed repair of costumes (CM 4)*** The costume master/mistress watches rehearsal(s) and take(s) closely for the way actor(s) and actress(es) use and abuse the costumes and prepares for any needed repair of the costumes between rehearsal(s) and take(s).

***Making adjustments and repairs to costumes (CM 5)*** The costume master/mistress monitors the use of the costumes and after rehearsal(s) and take(s) is ready to make adjustments and repairs.

***Knowing the rehearsal and take schedule for the actor(s) and actress(es) and assisting in readying them for subsequent rehearsal(s) and take(s) (MA 7)*** The make-up artist knows the rehearsal and take schedule for the actor(s) and actress(es) so as to assist them for subsequent rehearsal(s) and take(s).

***Checking with the continuity person between script line units and takes (T 8)*** The cast always checks with the continuity person for reestablishing details of dialogue, costume, make-up, and properties before beginning a subsequent unit or take. The actor(s) and actress(es) have to begin subsequent takes with the same details they ended a previous take to facilitate a postproduction edit required at that point.

***Making frequent checks with the costume master/mistress and the make-up artist to review and repair costumes and make-up when necessary (T 9)*** Actor(s) and actress(es) should make a habit of checking frequently with the costume master/mistress and the make-up artist for costume and make-up review. If repair of either is needed it should be made as soon as possible.

***Preparing for the production of any still photographs or production shots (P 5)*** The producer may need to prepare for the photographing of any still photographs and shots of the product that may be needed from the set after videotaping is complete. Still photographs may be used for other advertising media outside the video commercial. A product shot—the product alone on the set—may be required for the video commercial also. Before a studio wrap is called these additional needs must be met.

***Taking required shots; for example, still or product shots (D 16)*** The director follows the producer's instructions in videotaping product shots or instructing the studio to prepare for the photography of any other still photographs required of the set, cast, or crew before a studio wrap.

***Removing old spike marks before a new script line unit blocking begins (F 7)*** The floor director begins to remove old spike marks from the studio floor when the director calls a wrap for a previous script line unit.

***Overseeing the director's call for a take, a retake, next script line unit, a studio wrap, or a studio strike (F 8)*** The floor director oversees the studio as the director makes a call on a previous take. If the director wants another take, the floor director readies everyone in the studio for the beginning of that take. If the director calls for a wrap and the next script line unit, the floor director prepares for that unit by announcing the director's intentions. If the director calls a studio strike, then the floor director oversees the striking of the studio and studio personnel.

***Responding to the director's call, whether a retake, a wrap, the next script line unit, or a studio strike (TD 7)*** The technical director is ready for whatever the director's next call may be, whether a retake of the previous unit, a wrap for the previous unit, the intention to begin the next scheduled unit, or a studio strike.

***Reestablishing the details of a previous take before succeeding take(s) or unit(s) (CP 5)*** If the director's call is to retake a unit, the continuity person instructs the actor(s) and actress(es) on the details of dialogue, costume, properties, and make-up before the beginning of the previous take. These details ensure against unmatched edits in postproduction.

***Announcing production intentions (D 17)***   The director makes a production decision and announces that decision to the cast and crew. The director could choose to do another take, call a wrap for a particular script line unit and the intention to block the next scheduled unit, or call a studio strike. The director makes production intentions clear to the assistant director and technical director who relay them to those crew and cast subordinate to them or on their intercom networks.

***Securing talent release signatures (P 6)***   The producer secures any required talent release signatures of the cast. If studio extras were employed for the production session, talent release signatures are secured from each one before they depart.

***Preparing to block the next script line unit or striking the cameras (CO 13)***   The camera operators prepare to begin the process of studio production again with the blocking of the next scheduled script line unit or they strike the cameras by capping the lenses; locking pan, tilt, and pedestal; and recoiling camera cables.

***Preparing to light the next script line unit or set or striking floor lights and turning off set lights at a strike call (LD 7)***   The lighting director prepares to light the next script line unit set or begins to strike the floor lights and turns off the set lights at the director's strike call.

***Preparing for the next script line unit with a wrap call by the director or disassembling microphones, holders, and cables and closing down the audio control board with a strike call by the director (A 16)***   The audio director prepares to continue sound coverage with the director's call for a wrap on one script line unit and progress to another. The audio director strikes the microphones, holders, and cables if the director calls a studio strike.

***Preparing for the next script line unit with a wrap call by the director or disassembling microphone(s), boom(s), and cables with a studio strike call from the director (MBG/O 13)***   The microphone boom grip(s)/ operator(s) prepare to continue sound coverage with a wrap call for one script line unit and the announcement of the next unit. If the director calls for a studio strike, the grip(s)/operator(s) begin disassembling microphone(s), boom(s), and cables.

***Capping the cameras with the director's call for a wrap and strike (VEG 8)***   The video engineer responds to the director's call for a studio wrap and strike by electronically capping the cameras from camera control.

***Overseeing striking the control room and securing the director's master script and videotape log form (AD 10)***   The assistant director conveys to the crew the director's call to strike, oversees the strike of the control room, and secures the director's master script and videotape log for the next production session.

***Overseeing the complete strike of the production studio (F 9)***   The floor director oversees the complete strike of the studio, which entails monitoring the storage of cameras and breakdown of studio hardware, shutting down the set, sorting properties, and, in general, cleaning up after the crew and cast.

***Shutting down the switcher (TD 8)***   The technical director shuts the switcher down with the director's wrap call.

***Rewinding source videotape stock and accurately labeling the content of the videotape on the tape stock and storage case with the director's call for a studio strike (VTRO 10)***   When the director calls for a strike, the videotape recorder operator rewinds the source tape and shuts off the record deck. The videotape recorder operator then accurately labels both the videotape stock and its storage case with the content of the session's tapings.

***Shutting down the character generator and securing memory record of content (PA 5)***   The production assistant shuts the character generator down with the director's call for a strike.

***Reordering continuity forms (CP 6)***   The continuity person reorders all of the continuity forms recorded for the production session in the order of takes.

***Collecting, cleaning, and storing properties after take(s) and studio strike call (PM 6)***   The properties master/mistress collects properties from the cast and the set for cleaning if needed and storing or return.

***Retrieving, repairing, and cleaning costumes used during production (CM 6)***   The costume master/mistress retrieves costumes from the actor(s) and actress(es) after the studio strike is called. Some costumes will have to be repaired and some cleaned before being returned. Make-up residue is always a problem for costumes, so costumes require regular cleaning.

***Assisting actor(s) and actress(es) in removing make-up and cleaning up after a strike (MA 8)***   The make-up artist assists actor(s) and actress(es) in removing their make-up. Detailed make-up (e.g., for aging or facial hair) requires the assistance of the make-up artist, especially if elements of the make-up are to be reused. Of all areas of production, make-up requires the most effort to clean up not only the cast but also the facilities.

***Removing costuming and make-up and cleaning up (T 10)***   Actor(s) and actress(es) remove costumes and make-up after the director's call for a strike. The costume master/mistress and make-up artist assist the cast. The cast should report costume repair and cleaning needs to the costume master/mistress. Cleaning up costume and make-up facilities is a shared responsibility among the cast and the staff.

## THE POSTPRODUCTION PROCESS

The nature of some commercial productions is such that no postproduction work on the source videotape is necessary or expected. Some commercials run "straight through" production from beginning to end. The correct academy leader slate is added during studio production, the videotaped units taken repeatedly until the correct time length is achieved, and all character generator copy and artwork are videotaped and inserted correctly. Many commercials leave the studio completed after a production session.

The postproduction process for other commercials is an important step in the production of the commercial. The production of the television studio commercial was accomplished by breaking a commercial script into script line units. These units, videotaped during the production process, now must be edited together into the completed commercial.

At the beginning of postproduction, the producer and director have the source videotape(s) from the videotape recorder operator, labeled and ordered by script line unit. These source tapes and adequate master videotape stock have been striped by the videotape recorder operator with SMPTE time code. The director and producer have the videotape log that was kept by the assistant director during studio production. These videotape log forms list all videotape takes of all script line units. The director also has the master script copy, updated during studio production and reordered into the order of the original commercial. The audio director supplies any prerecorded audio tracks, remaining music, or sound effects not already wedded to the source videotapes. Postproduction can begin.

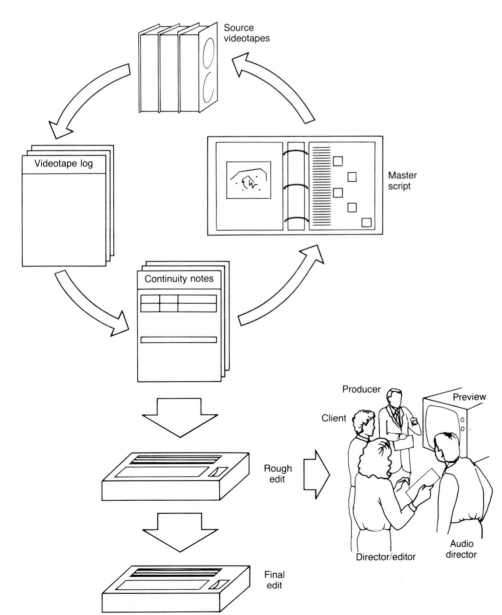

**FIGURE 2–30**
**Commercial postproduction.** Some commercials produced in the studio will require postproduction. Five steps include preparing for postproduction, gathering all resources, performing a rough edit, reviewing the edit with the client, and performing the final edit.

### • Personnel

Personnel required for the postproduction of the commercial will vary by the custom of a facility and the skills of the postproduction personnel. The producer continues in the role defined for the producer in production. The director may choose to do the postproduction editing. In many ways, the best person to do the postproduction editing is the director because the director designed the production and directed the studio session. On the other hand, it would not be uncommon for a technical director/editor to serve as a videotape editor. If a technical director/editor is going to do the editing, the producer and director will work with the editor during the editing sessions.

*Producer (P)*   The producer's role from the preproduction and production stages continues into postproduction. The role continues to be a supervisory and organizing role. The producer attends to scheduling and creative consulting before and during editing.

*Director (D)*   The director continues into postproduction as the catalyst in seeing the commercial through to completion. In some operations, the director will do the editing. In others, a technical director/editor will do the technical editing under the guidance and creative direction of the director.

*Technical director/editor (TD/E)*   When the director of the production does not do the actual editing of the source tapes, a technical director/editor will be called upon to do it. This role is generally a button-pushing role under the guidance and creative direction of the director with assistance from the producer.

*Audio director (A)*   The audio director plays a role in postproduction insofar as the sound track of the final edited master tape may need additional postproduction work. Primarily, the prerecorded audio track, music, and sound effects track will have to be laid down on the master videotape and mixed onto a single audio track on the master tape during postproduction. While a technical director/editor has the skills to do this, the audio director will have to supply the audio sources. When the video portion of the master tape is completely edited, the audio director may be required to separate the dialogue track and sound effects track from the master videotape, sweeten it, and mix the remaining music onto the master tape.

*Assistant director (AD)*   The assistant director may be required to continue to assist the director during postproduction. The assistant director was very close to the director's master script during production and logged all videotape takes. This experience can become an invaluable resource to the producer and director (and technical director/editor) during postproduction.

*Videotape recorder operator (VTRO)*   The videotape recorder operator plays a limited but important role in postproduction. The videotape recorder operator possesses the source videotapes, has labeled them after production, and has striped the source videotapes with SMPTE time code. The master videotape stock to be used in postproduction should also be striped with time code.

### • Postproduction Stages

*Arranging the postproduction editing schedule and the facilities (P 1)*   The producer begins postproduction by arranging for the editing facilities and scheduling editing time. The schedule and available editing facilities will have to account for the videotape editor. The director may edit or a technical director/editor will do the editing.

*Supervising postproduction preparation (P 2)*   The producer continues in postproduction by supervising preparation for editing. This means that resources, including source and master videotapes, have to be secured, the master script reordered, the videotape take log secured, postproduction cue sheet form completed, audio tracks secured, and previewing appointments made with the client or agency to view the rough edit of the commercial.

*Supervising postproduction preparation (D 1)*   The director shares in supervising postproduction preparation with the producer. Details of the production required for postproduction were under the direction of the director, including the master script, the videotape log, and audio tracks.

*Reordering and editing the master script (D 2)*   The director controls the primary postproduction resource (after the source videotapes)—the master script. The master script (depending on its length) may require some reordering into the original order of the commercial. During blocking and taping the commercial was broken into script line units and produced in the studio in perhaps a different order from the original order of the script. The master script must be reordered for postproduction. There may be other editing decisions to be made in terms of production problems or script interpretation that occurred during preproduction and studio production that have to be decided before postproduction and then corrected (e.g., by using cut-aways (pickup shots) or prerecorded exterior B-rolls).

*Providing prerecorded tracks, music tracks, and sound effects audio sources (A 1)*   The audio director is responsible for providing the prerecorded audio tracks, music tracks, or sound effects tracks to the postproduction editing session. Some prerecorded audio, music, and sound effects tracks may already be a part of the source tapes. Other music and effects may be needed across edits to be relaid in postproduction. (See the effects/music cue sheet form.)

*Turning the videotape take log forms over to the producer (AD 1)*   The assistant director turns over to the producer for postproduction reference the videotape log forms. The videotape logs are a detailed resource of the order and disposition of each take on the source tapes.

*Creating the postproduction cue sheet form (D 3)*   The director prepares the postproduction cue sheet form as a prelude to editing. The postproduction cue sheet form collects necessary information from the master script and the videotape log forms into a coherent order to facilitate postproduction editing. The thoroughness of this form will permit a person (e.g., a technical director/editor) unfamiliar with the production the chance to cut a master edit with ease. (See the postproduction cue sheet form.)

### Striping the source and master videotapes (VTRO 1)

The videotape recorder operator stripes all videotape stock with SMPTE time code in preparation for editing. The time code permits an accurate determination of edit points as videotape editing is performed from the source tapes to the master tape during postproduction.

### Turning the source tapes and the master videotape stock over to the producer (VTRO 2)

The videotape recorder operator turns all videotapes over to the producer. The videotapes include the source tapes recorded during studio production and the videotape stock necessary for the master videotape.

### Editing a rough cut edition of the commercial (D 4)

The director begins a rough cut editing of the commercial. A rough cut serves a number of purposes. First, a rough cut permits the director to sense the flow and pacing of dialogue and effects across edits. As individual script line units of the commercial are produced it becomes difficult or impossible to capture a sense of the whole commercial. A rough edit reveals where editing can control pacing: More rapid editing can pick up the pacing and fewer edits slow the pacing. A rough edit will also reveal where tighter edits will serve the dialogue and pacing. Shaving frames off at edit points will create a sharper perception of multiple camera production.

Second, the time length of the commercial is important. The commercial has to hit its contracted length of 10, 15, 20, 30, or 60 seconds exactly. A rough cut allows an evaluation of length requirements and areas that can be lengthened or shortened.

Third, a rough edit gives the director a sense of the whole, something not yet experienced. The rough edit becomes an essential point of reference for a final editing.

And, fourth, before a final edit, the client or agency should see the rough edit. Their sense of what they wanted and what they approved versus what was produced enters into the final edit decisions.

### Observing the editing session with the director and assisting in edit decision making (P 3)

The producer keeps the editing schedule with the director and becomes a partner with the director in the editing decisions. After the director, the producer originated the creative design of the commercial, knows what the client or agency approved, and has much to share in editing decisions. The final decision should remain with the director at this stage.

### Assisting the director during postproduction (AD 2)

The assistant director is of service to the director during postproduction. A major task during editing is to search out units from the source videotapes for editing or to review other takes. The assistant director can help the

director (or technical director/editor) and save editing time by searching out source tapes for takes while editing continues.

### Editing from the master script, the videotape log forms, and the postproduction cue sheet form under the supervision of the producer and director (TD/E 1)

If the technical director/editor does the editing, the technical director/editor edits from the master script, the videotape log forms, and the postproduction cue sheet form. The first editing is a rough edit. The technical director/editor works under the supervision of the producer and director.

### Reviewing the rough edit (P 4)

The producer reviews the rough edit with the director as the opportunity to recommend changes to the final edit, placement of edits, cut-aways, music, and effects. This is the first opportunity to experience the commercial as a whole since reading the script.

### Reviewing the rough edit (D 5)

The director reviews the rough edit with the producer as a step to making decisions affecting the final edit. Edit pacing has to be evaluated, music and sound effects planned, additional titling and credits created, and broadcast time block created. This step is a time for copious note taking as a prelude to undertaking a final edit.

### Previewing the rough edit with the client or agency (P 5)

The producer reviews the rough edit with the client or agency with whom the producer was working. This is the first opportunity that the client or agency will have to see the commercial since approving the script and storyboard. The creative interpretation of a director has since entered into the production of the commercial. Before a final edit, the client or agency must provide input into the final edited commercial. The director previews the rough edit with the client or agency to offer any justification for changes or interpretation during production and editing.

### Previewing the rough edit with the client or agency (D 6)

It is important for the director to attend the preview of the rough edit of the commercial with the producer and the client or agency. It is the director who appropriately takes creative and interpretive liberty with the approved script and storyboard during production. At this time the director can observe the reaction of the client or agency to the rough edit; justify, where necessary, any changes the director made to the script or storyboard; perhaps argue a creative or interpretive point with the client or agency; and make notes for a final edit.

### Assisting the director and technical director/editor with audio problems and additions to the final edit (A 2)

The audio director assists the director or technical director/editor with any decisions involving the audio portion of the commercial. Given the expertise of the audio director, some questions about the sound track are more easily solved with the input of the audio director. The audio director reviews the rough edit with the director.

*Making the final edit. (D 7)*   The director begins the final edit after reviewing the rough edit with the producer, audio director, and client or agency. The final edit attempts to embody all suggestions and improvements generated by the meeting with the producer, audio director, and client or agency.

*Sweetening the sound track of the final edit (A 3)*
The audio director may be required to do postproduction sweetening to the entire sound track of the final edit as a final responsibility to the production. This entails separating the time coded sound track from the time coded video track. Once this is done, the sound track can be worked with (e.g., audio reproduction of the studio sound can be cleaned up and any prerecorded audio tracks, music tracks, and sound effects can be added). The sound track is then rewedded to the video track with accurate time coding.

*Adding music and sound effects and mixing the audio channels (TD/E 2)*   If a technical director/editor is editing the commercial, after the final video editing the technical director/editor adds music and sound effects to the second audio channel and mixes the audio channels to a single channel. The audio director may be required to sweeten the audio track with more polished audio postproduction.

*Labeling the edited master videotape and removing the record button (D 8)*   When final video editing is completed and audio sweetening done, the director labels the edited master and removes the record button to prevent accidental erasing of the videotape.

*Labeling the edited master videotape and removing the record button (TD/E 3)*   If the editing was done by the technical director/editor, the technical director/editor labels the final edited master and removes the record button to prevent accidental erasing of the videotape.

*Turning the edited master videotape over to the producer (TD/E 4)*   The technical director/editor turns the final edited master videotape over to the producer.

*Turning the edited master videotape over to the producer (D 9)*   The director turns the final edited master videotape over to the producer.

*Approving the final edited master videotape (P 6)*
The producer has the final approval of the edited master videotape. This may require a final previewing with the director or technical director/editor.

*Turning the final edited master videotape over to the client or agency (P 7)*   The producer turns the final edited master videotape over to the client or agency for final viewing, approval, and acceptance.

# Chapter 3

# Commercial Production Organizing Forms

There are a diversity and breadth of roles and task responsibilities in television studio commercial producing and production unknown in other television genres. This very diversity and breadth warrant some technique within which role and task responsibility of personnel can be successfully met. It is frequently said in television production that the success of a production is a function of the extent and degree of preproduction.

Organizing forms are presented in this chapter for almost all information gathering and task preparation stages required for studio commercial preproduction and production. The majority of the forms serve to design and order commercial producing tasks. The remaining forms help to prepare for the actual studio production of the commercial script. Included are forms that were designed to assist in the design of commercial production and in the development of the cast from audition to characteri-

zation. As with many resources in a creative and technological medium, not all forms will be equally functional. These forms should be used when they facilitate the tasks for which they were created. They should not become ends in themselves. Forms that serve the production should be used and perhaps adapted to the requirements of a particular commercial script or studio production facility. Still others may be needed and may serve a one time only function. Others may not be required and should be overlooked.

The forms presented here are designed to be used by producing and production personnel for particular role accomplishment. Those roles are listed in the upper portions of the respective forms. These forms can be removed or photocopied. Note that each form's structure and function is detailed in Chapter 4.

# MARKETING ANALYSIS
## MULTIPLE CAMERA VIDEO COMMERCIAL PRODUCTION

Producer:                          Commercial Title:
Client Representative:             Client:
Date:    /    /                    Product/Service:

## GOAL(S) AND OBJECTIVE(S) FOR THIS COMMERCIAL

List each and every goal and objective, even the most obvious, intended with this commercial.

## TARGET AUDIENCE FOR THIS COMMERCIAL

Describe who uses this product or service and who purchases this product or service, by age, sex, occupation, economic status, etc.

Suggested audio/video production values related to the characteristics of the target audience:

## PRODUCTION STATEMENT

Write a one sentence statement that will be kept in mind during every moment of production.

# PRODUCTION BUDGET
## MULTIPLE CAMERA VIDEO COMMERCIAL PRODUCTION

Producer:                                    Commercial Title:
Director:                                     Client:
Date:    /    /                               Commercial Length:      :
No. Preproduction Days: ☐  Hours: ☐           Production Date:    /    /
No. Studio Production Days: ☐  Hours: ☐       Commercial Completion Date:    /    /
No. Postproduction Days: ☐  Hours: ☐

| SUMMARY OF PRODUCTION COSTS | ESTIMATED | ACTUAL |
|---|---|---|
| 1. Rights and Clearances | | |
| 2. Commercial Script | | |
| 3. Producer and Staff | | |
| 4. Director and Staff | | |
| 5. Talent | | |
| 6. Benefits | | |
| 7. Production Facility | | |
| 8. Production Staff | | |
| 9. Camera, Videotape Recording, Video Engineering | | |
| 10. Set Design, Construction, Decoration | | |
| 11. Set Lighting | | |
| 12. Properties | | |
| 13. Costuming | | |
| 14. Make-up and Hairstyling | | |
| 15. Audio Production | | |
| 16. Videotape and Film Stock, Film Processing | | |
| 17. Postproduction Editing | | |
| 18. Music | | |
| 19. Artwork and Photography | | |
| 20.                              Subtotal | | |
| 21. Contingency | | |
|                                  Grand Total | | |

COMMENTS

## RIGHTS AND CLEARANCES

| DESCRIPTION | ESTIMATED | ACTUAL |
|---|---|---|
| 1. Script Rights Purchase | | |
| 2. Synchronization Rights | | |
| 3. Logo Registration | | |
| 4. Other Rights | | |
| 5. Music Clearances | | |
| 6. | | |
| Subtotal | | |

## COMMERCIAL SCRIPT

| DESCRIPTION | ESTIMATED | ACTUAL |
|---|---|---|
| 7. Writer Salary: No. ( ) @ ( ) | | |
| 8. | | |
| 9. Script Editor Salary: No. ( ) @ ( ) | | |
| 10. | | |
| 11. Secretary Salary: No. ( ) @ ( ) | | |
| 12. Searches (claims, libel avoidance, etc.) | | |
| 13. Office Supplies | | |
| 14. Photocopying | | |
| 15. Telephone | | |
| 16. Postage | | |
| 17. Miscellaneous | | |
| Subtotal | | |

## PRODUCER AND STAFF

| CREW | ESTIMATED | | | | ACTUAL | | | |
|---|---|---|---|---|---|---|---|---|
| | Days | Rate | O/T Hrs | Total | Days | Rate | O/T Hrs | Total |
| 18. Producer | | | | | | | | |
| 19. Preproduction | | | | | | | | |
| 20. Production | | | | | | | | |
| 21. Postproduction | | | | | | | | |
| 22. Secretary: No. ( ) | | | | | | | | |
| 23. | | | | | | | | |
| 24. Producer Supplies | | | | | | | | |
| 25. Photocopying | | | | | | | | |
| 26. Telephone | | | | | | | | |
| 27. Postage | | | | | | | | |
| 28. Travel Expenses | | | | | | | | |
| 29. Per Diem | | | | | | | | |
| 30. Miscellaneous | | | | | | | | |
| 31. | | | | | | | | |
| Subtotal | | | | | | | | |

## DIRECTOR AND STAFF

| CREW | ESTIMATED | | | | ACTUAL | | | |
|---|---|---|---|---|---|---|---|---|
| | Days | Rate | O/T Hrs | Total | Days | Rate | O/T Hrs | Total |
| 32. Director | | | | | | | | |
| 33.　　　　Preproduction | | | | | | | | |
| 34.　　　　Production | | | | | | | | |
| 35.　　　　Postproduction | | | | | | | | |
| 36. Assistant Director | | | | | | | | |
| 37.　　　　Preproduction | | | | | | | | |
| 38.　　　　Production | | | | | | | | |
| 39.　　　　Postproduction | | | | | | | | |
| 40. Secretary: No. (　　) | | | | | | | | |
| 41. | | | | | | | | |
| 42. Director's Supplies | | | | | | | | |
| 43.　Photocopying | | | | | | | | |
| 44.　Telephone | | | | | | | | |
| 45.　Postage | | | | | | | | |
| 46. Expenses | | | | | | | | |
| 47. Miscellaneous | | | | | | | | |
| 48. | | | | | | | | |
| | Subtotal | | | | | | | |

## TALENT

| DESCRIPTION | ESTIMATED | | | | ACTUAL | | | |
|---|---|---|---|---|---|---|---|---|
| | Days | Rate | O/T Hrs | Total | Days | Rate | O/T Hrs | Total |
| 49.　Cast | | | | | | | | |
| 50.　CHARACTER　│　ACTOR | | | | | | | | |
| 51. | | | | | | | | |
| 52. | | | | | | | | |
| 53. | | | | | | | | |
| 54. | | | | | | | | |
| 55. | | | | | | | | |
| 56. | | | | | | | | |
| 57. | | | | | | | | |
| 58. | | | | | | | | |
| 59.　Extras　　　No.(　　) | | | | | | | | |
| 60. | | | | | | | | |
| | Subtotal | | | | | | | |

## BENEFITS

| DESCRIPTION | TOTAL |
|---|---|
| 61.　Insurance Coverage | |
| 62.　Health Plan | |
| 63.　Pension Plan (etc.) | |
| 64.　Taxes (FICA, etc.) | |
| 65.　Equity/Guild/Union Costs | |
| | |
| Subtotal | |

## PRODUCTION FACILITY

| SPACE | ESTIMATED | | | | ACTUAL | | | |
|---|---|---|---|---|---|---|---|---|
| | Days | Rate | O/T Hrs | Total | Days | Rate | O/T Hrs | Total |
| 66. Production Studio | | | | | | | | |
| 67. Studio Camera No.( ) | | | | | | | | |
| 68. Teleprompter | | | | | | | | |
| 69. Floor Monitor No.( ) | | | | | | | | |
| 70. | | | | | | | | |
| 71. Control Room | | | | | | | | |
| 72. Production Switcher | | | | | | | | |
| 73. Digital Video Effects | | | | | | | | |
| 74. Still Store | | | | | | | | |
| 75. Intercom Network | | | | | | | | |
| 76. Character Generator | | | | | | | | |
| 77. | | | | | | | | |
| 78. Audio Control Board | | | | | | | | |
| 79. Cartridge Playback | | | | | | | | |
| 80. Cassette Playback | | | | | | | | |
| 81. Reel-to-Reel Playback | | | | | | | | |
| 82. Turntable | | | | | | | | |
| 83. Compact Disc Player | | | | | | | | |
| 84. Studio Foldback | | | | | | | | |
| 85. | | | | | | | | |
| 86. Master Control | | | | | | | | |
| 87. Record Deck No.( ) | | | | | | | | |
| 88. Playback Deck No.( ) | | | | | | | | |
| 89. Other | | | | | | | | |
| 90. | | | | | | | | |
| 91. Dressing Rooms | | | | | | | | |
| 92. Costume Storage Room | | | | | | | | |
| 93. Make-Up Facility | | | | | | | | |
| 94. Green Room | | | | | | | | |
| 95. Other | | | | | | | | |
| | Subtotal | | | | | | | |

## PRODUCTION STAFF

| STAFF | ESTIMATED | | | | ACTUAL | | | |
|---|---|---|---|---|---|---|---|---|
| | Days | Rate | O/T Hrs | Total | Days | Rate | O/T Hrs | Total |
| 96. Technical Director | | | | | | | | |
| 97. Preproduction | | | | | | | | |
| 98. Production | | | | | | | | |
| 99. Postproduction | | | | | | | | |
| 100. Floor Director | | | | | | | | |
| 101. Preproduction | | | | | | | | |
| 102. Production | | | | | | | | |
| 103. Postproduction | | | | | | | | |
| 104. Production Assistant | | | | | | | | |
| 105. Preproduction | | | | | | | | |
| 106. Production | | | | | | | | |
| 107. Postproduction | | | | | | | | |

## PRODUCTION STAFF (Continued)

| STAFF | ESTIMATED | | | | ACTUAL | | | |
|---|---|---|---|---|---|---|---|---|
| | Days | Rate | O/T Hrs | Total | Days | Rate | O/T Hrs | Total |
| 108. Continuity Person | | | | | | | | |
| 109.     Preproduction | | | | | | | | |
| 110.     Production | | | | | | | | |
| 111.     Postproduction | | | | | | | | |
| 112. Other Crew | | | | | | | | |
| 113. | | | | | | | | |
| 114. Miscellaneous | | | | | | | | |
| 115. | | | | | | | | |
| | Subtotal | | | | | | | |

## CAMERAS, VIDEOTAPE RECORDING, AND VIDEO ENGINEERING

| DESCRIPTION | ESTIMATED | | | | ACTUAL | | | |
|---|---|---|---|---|---|---|---|---|
| | Days | Rate | O/T Hrs | Total | Days | Rate | O/T Hrs | Total |
| 116. Camera Operator No. ( ) | | | | | | | | |
| 117.     Preproduction | | | | | | | | |
| 118.     Production | | | | | | | | |
| 119.     Postproduction | | | | | | | | |
| 120. VTR Operator | | | | | | | | |
| 121.     Preproduction | | | | | | | | |
| 122.     Production | | | | | | | | |
| 123.     Postproduction | | | | | | | | |
| 124. Video Engineer | | | | | | | | |
| 125.     Preproduction | | | | | | | | |
| 126.     Production | | | | | | | | |
| 127.     Postproduction | | | | | | | | |
| 128. Equipment | | | | | | | | |
| 129. Rental | | | | | | | | |
| 130. Purchases | | | | | | | | |
| 131. Maintenance/Repair | | | | | | | | |
| 132. | | | | | | | | |
| 133. Miscellaneous | | | | | | | | |
| 134. | | | | | | | | |
| | Subtotal | | | | | | | |

## SET DESIGN, CONSTRUCTION, AND DECORATION

| DESCRIPTION | ESTIMATED | | | | ACTUAL | | | |
|---|---|---|---|---|---|---|---|---|
| | Days | Rate | O/T Hrs | Total | Days | Rate | O/T Hrs | Total |
| 135. Set Designer | | | | | | | | |
| 136. | | | | | | | | |
| 137. Set Construction | | | | | | | | |
| 138. Materials | | | | | | | | |
| 139. Labor | | | | | | | | |
| 140. | | | | | | | | |
| 141. Miscellaneous | | | | | | | | |
| 142. | | | | | | | | |
| 143. Set Dressing Labor | | | | | | | | |
| 144. | | | | | | | | |

## SET DESIGN, CONSTRUCTION, AND DECORATION (Continued)

| DESCRIPTION | ESTIMATED | | | | ACTUAL | | | |
|---|---|---|---|---|---|---|---|---|
| | Days | Rate | O/T Hrs | Total | Days | Rate | O/T Hrs | Total |
| 145. Set Dressing Props | | | | | | | | |
| 146.  Purchases | | | | | | | | |
| 147.  Rentals | | | | | | | | |
| 148.  Cleaning/Loss/Damage | | | | | | | | |
| 149. | | | | | | | | |
| 150. Miscellaneous | | | | | | | | |
| 151. | | | | | | | | |
| Subtotal | | | | | | | | |

## PROPERTIES

| DESCRIPTION | ESTIMATED | | | | ACTUAL | | | |
|---|---|---|---|---|---|---|---|---|
| | Days | Rate | O/T Hrs | Total | Days | Rate | O/T Hrs | Total |
| 152. Props Master/Mistress | | | | | | | | |
| 153. | | | | | | | | |
| 154.  Property Labor | | | | | | | | |
| 155.  Vehicle | | | | | | | | |
| 156.  Transportation | | | | | | | | |
| 157.  Storage | | | | | | | | |
| 158. | | | | | | | | |
| 159. Animal(s)    No.( ) | | | | | | | | |
| 160.  Animal Handler | | | | | | | | |
| 161.  Feed/Stabling | | | | | | | | |
| 162. | | | | | | | | |
| 163. Action Properties | | | | | | | | |
| 164.  Rental | | | | | | | | |
| 165.  Purchases | | | | | | | | |
| 166.  Cleaning/Loss/Damage | | | | | | | | |
| 167. | | | | | | | | |
| 168. Hand Properties | | | | | | | | |
| 169.  Rental | | | | | | | | |
| 170.  Purchases | | | | | | | | |
| 171.  Storage | | | | | | | | |
| 172.  Food | | | | | | | | |
| 173. | | | | | | | | |
| 174. Miscellaneous | | | | | | | | |
| Subtotal | | | | | | | | |

## SET LIGHTING

| DESCRIPTION | ESTIMATED | | | | ACTUAL | | | |
|---|---|---|---|---|---|---|---|---|
| | Days | Rate | O/T Hrs | Total | Days | Rate | O/T Hrs | Total |
| 175. Lighting Director | | | | | | | | |
| 176.    Preproduction | | | | | | | | |
| 177.    Production | | | | | | | | |
| 178.    Postproduction | | | | | | | | |
| 179. Expendables | | | | | | | | |
| 180.   (gels, etc.) | | | | | | | | |
| 181. Lighting Equipment | | | | | | | | |
| 182.  Rental | | | | | | | | |
| 183.  Purchases | | | | | | | | |

## SET LIGHTING (Continued)

| DESCRIPTION | ESTIMATED | | | | ACTUAL | | | |
|---|---|---|---|---|---|---|---|---|
| | Days | Rate | O/T Hrs | Total | Days | Rate | O/T Hrs | Total |
| 184. Light Instruments | | | | | | | | |
| 185. 1 kw Spot No.( ) | | | | | | | | |
| 186. 2 kw Spot No.( ) | | | | | | | | |
| 187.   kw Spot No.( ) | | | | | | | | |
| 188. 1 kw Scoop No.( ) | | | | | | | | |
| 189. 1 1/2 kw Scoop No.( ) | | | | | | | | |
| 190. 2 kw Scoop No.( ) | | | | | | | | |
| 191.   kw Scoop No.( ) | | | | | | | | |
| 192. Ellipsoidal Spotlight | | | | | | | | |
| 193. Follow Spot | | | | | | | | |
| 194. Broad/Softlight No.( ) | | | | | | | | |
| 195. Strip/Cyc Light No.( ) | | | | | | | | |
| 196. Miscellaneous | | | | | | | | |
| 197. | | | | | | | | |
| | Subtotal | | | | | | | |

## AUDIO PRODUCTION

| DESCRIPTION | ESTIMATED | | | | ACTUAL | | | |
|---|---|---|---|---|---|---|---|---|
| | Days | Rate | O/T Hrs | Total | Days | Rate | O/T Hrs | Total |
| 198. Audio Director | | | | | | | | |
| 199.      Preproduction | | | | | | | | |
| 200.      Production | | | | | | | | |
| 201.      Postproduction | | | | | | | | |
| 202. Mike Grip     No.( ) | | | | | | | | |
| 203.      Preproduction | | | | | | | | |
| 204.      Production | | | | | | | | |
| 205.      Postproduction | | | | | | | | |
| 206. Boom Operator No.( ) | | | | | | | | |
| 207.      Preproduction | | | | | | | | |
| 208.      Production | | | | | | | | |
| 209.      Postproduction | | | | | | | | |
| 210. Audio Boom    No.( ) | | | | | | | | |
| 211. Fishpole     No.( ) | | | | | | | | |
| 212. Audio Equipment | | | | | | | | |
| 213.   Rental | | | | | | | | |
| 214.   Purchases | | | | | | | | |
| 215. Prerecording | | | | | | | | |
| 216.   Narrator | | | | | | | | |
| 217.   VO Narration | | | | | | | | |
| 218. Microphones | | | | | | | | |
| 219.   Shotgun     No.( ) | | | | | | | | |
| 220.   Lavaliere    No.( ) | | | | | | | | |
| 221.   Handheld    No.( ) | | | | | | | | |
| 222.   Boom       No.( ) | | | | | | | | |
| 223.   Wireless    No.( ) | | | | | | | | |
| 224.   Desk       No.( ) | | | | | | | | |
| 225.   Stand      No.( ) | | | | | | | | |
| 226. | | | | | | | | |
| 227. Miscellaneous | | | | | | | | |
| | Subtotal | | | | | | | |

## COSTUMING

| DESCRIPTION | ESTIMATED | | | | ACTUAL | | | |
|---|---|---|---|---|---|---|---|---|
| | Days | Rate | O/T Hrs | Total | Days | Rate | O/T Hrs | Total |
| 228. Costumer | | | | | | | | |
| 229.     Preproduction | | | | | | | | |
| 230.     Production | | | | | | | | |
| 231.     Postproduction | | | | | | | | |
| 232. Costumes | | | | | | | | |
| 233.  Rental | | | | | | | | |
| 234.  Purchases | | | | | | | | |
| 235.  Loss/Damage | | | | | | | | |
| 236.  Cleaning | | | | | | | | |
| 237. Costume Building | | | | | | | | |
| 238.  Materials | | | | | | | | |
| 239. Costume Labor | | | | | | | | |
| 240. Miscellaneous | | | | | | | | |
| 241. | | | | | | | | |
| | Subtotal | | | | | | | |

## MAKE-UP AND HAIRSTYLING

| DESCRIPTION | ESTIMATED | | | | ACTUAL | | | |
|---|---|---|---|---|---|---|---|---|
| | Days | Rate | O/T Hrs | Total | Days | Rate | O/T Hrs | Total |
| 242. Make-up Artist | | | | | | | | |
| 243.  Make-up Supplies | | | | | | | | |
| 244. | | | | | | | | |
| 245. Hairstylist | | | | | | | | |
| 246.  Hairstyling Supplies | | | | | | | | |
| 247. Wig | | | | | | | | |
| 248.  Rental | | | | | | | | |
| 249.  Purchases | | | | | | | | |
| 250. Miscellaneous | | | | | | | | |
| 251. | | | | | | | | |
| | Subtotal | | | | | | | |

## VIDEOTAPE AND FILM STOCK, FILM PROCESSING

| DESCRIPTION | ESTIMATED | | | | ACTUAL | | | |
|---|---|---|---|---|---|---|---|---|
| | Days | Rate | O/T Hrs | Total | Days | Rate | O/T Hrs | Total |
| 252. Videotape: | | | | | | | | |
| 253.   1" ( : ) x ($ ) | | | | | | | | |
| 254.  3/4" (:20) x ($ ) | | | | | | | | |
| 255.  3/4" (:30) x ($ ) | | | | | | | | |
| 256.  3/4" (:60) x ($ ) | | | | | | | | |
| 257. Film Stock | | | | | | | | |
| 258.  35mm | | | | | | | | |
| 259. Film Processing | | | | | | | | |
| 260. Miscellaneous | | | | | | | | |
| 261. | | | | | | | | |
| | Subtotal | | | | | | | |

## POSTPRODUCTION EDITING

| DESCRIPTION | ESTIMATED | | | | ACTUAL | | | |
|---|---|---|---|---|---|---|---|---|
| | Days | Rate | O/T Hrs | Total | Days | Rate | O/T Hrs | Total |
| 262. Technical Director | | | | | | | | |
| 263. | | | | | | | | |
| 264. Editing Suite Rental | | | | | | | | |
| 265. Playback/Search Deck | | | | | | | | |
| 266. Master Tape Stock | | | | | | | | |
| 267. SMPTE Coding | | | | | | | | |
| 268. Stock Video Footage | | | | | | | | |
| 269. Audio Mixing | | | | | | | | |
| 270. Audio Sweetening Room | | | | | | | | |
| 271. Sound Effects | | | | | | | | |
| 272. Dubs ( ) x ($ )per | | | | | | | | |
| 273. Miscellaneous | | | | | | | | |
| 274. | | | | | | | | |
| | Subtotal | | | | | | | |

## MUSIC

| DESCRIPTION | ESTIMATED | | | | ACTUAL | | | |
|---|---|---|---|---|---|---|---|---|
| | Days | Rate | O/T Hrs | Total | Days | Rate | O/T Hrs | Total |
| 275. Music | | | | | | | | |
| 276. Purchases | | | | | | | | |
| 277. Royalties | | | | | | | | |
| 278. | | | | | | | | |
| 279. Song Writer | | | | | | | | |
| 280. Arranger | | | | | | | | |
| 281. Composer | | | | | | | | |
| 282. Conductor | | | | | | | | |
| 283. Vocalist No.( ) | | | | | | | | |
| 284. | | | | | | | | |
| 285. Recording Facility | | | | | | | | |
| 286. Recording Crew | | | | | | | | |
| 287. Audio Recording Tape | | | | | | | | |
| 288. | | | | | | | | |
| 289. Miscellaneous | | | | | | | | |
| 290. | | | | | | | | |
| | Subtotal | | | | | | | |

## ARTWORK AND PHOTOGRAPHY

| DESCRIPTION | ESTIMATED | | | | ACTUAL | | | |
|---|---|---|---|---|---|---|---|---|
| | Days | Rate | O/T Hrs | Total | Days | Rate | O/T Hrs | Total |
| 291. Artwork | | | | | | | | |
| 292. Animator | | | | | | | | |
| 293. | | | | | | | | |
| 294. Animation | | | | | | | | |
| 295. Materials | | | | | | | | |
| 296. Photography | | | | | | | | |
| 297. Processing | | | | | | | | |
| 298. | | | | | | | | |

## ARTWORK AND PHOTOGRAPHY (Continued)

| DESCRIPTION | ESTIMATED | | | | ACTUAL | | | |
|---|---|---|---|---|---|---|---|---|
| | Days | Rate | O/T Hrs | Total | Days | Rate | O/T Hrs | Total |
| 299. Photographer | | | | | | | | |
| 300. Still Photography | | | | | | | | |
| 301. Light Table | | | | | | | | |
| 302. Product Shot | | | | | | | | |
| 303. Set Shot | | | | | | | | |
| 304. | | | | | | | | |
| 305. Miscellaneous | | | | | | | | |
| 306. | | | | | | | | |
| | Subtotal | | | | | | | |

COMMENTS

# FACILITY REQUEST

## MULTIPLE CAMERA VIDEO COMMERCIAL PRODUCTION

**Production Facility:**

| | |
|---|---|
| Producer: | Commercial Title: |
| Director: | Client:      Length:   : |
| Date:   /   / | Production Dates:   /   /   to   /   / |

**Facility Space:**
- ☐ Studio
- ☐ Control Room
- ☐ Master Control
- ☐ Audio Control Room
- ☐ Make-Up Room
- ☐ Dressing Room
- ☐ Set Construction Workshop
- ☐ Costume Storage
- ☐ Property Storage

**Studio Requirements:**
- ☐ No. Studio Cameras
- ☐ Teleprompter
- ☐ No. Floor Monitors
-   Light Instruments
- ☐ _____
- ☐ Audio Boom Platform
- ☐ __kw spots   ☐ __kw scoops   ☐ ellipsoidal spotlight   ☐ follow spot   ☐ broad/softlight
- ☐ __kw spots   ☐ __kw scoops   ☐ strip/cyc lights   ☐ other_____

**Control Room Requirements:**      ☐ _____
- ☐ Production Switcher
- ☐ Character Generator
- ☐ Intercom Network: no. stations___
- ☐ DVE
- ☐ Still Store/E-men
- ☐ Light Control Board

**Master Control Area Requirements:**
- ☐ Videotape Decks
- ☐ Telecine
- ☐ _____
-   ☐ record machines
-   ☐ 35mm slides
-   ☐ playback machines
-   ☐ 16mm

**Audio Control Room Requirements:**
- ☐ Audio Control Board      ☐ Studio Foldback      ☐ _____
- ☐ Playback Input Units:
  - ☐ audio cartridge   ☐ audio cassette   ☐ reel-to-reel   ☐ turntable   ☐ compact disc
- ☐ Microphones:
  - ☐ lavaliere   ☐ handheld   ☐ boom   ☐ wireless   ☐ desk   ☐ stand

**Preproduction Requirements:**

| Set Construction/Decoration: | Date: / / | Hours: : to : |
|---|---|---|
| | Date: / / | Hours: : to : |
| | Date: / / | Hours: : to : |

| Set Lighting: | Date: / / | Hours: : to : |
|---|---|---|
| | Date: / / | Hours: : to : |
| | Date: / / | Hours: : to : |

Studio Production Personnel:

| Production Role | Personnel | Facility Crew | Program Crew |
|---|---|---|---|
| Director | | ☐ | ☐ |
| Assistant Director | | ☐ | ☐ |
| | | ☐ | ☐ |
| Production Assistant | | ☐ | ☐ |
| | | ☐ | ☐ |
| Technical Director | | ☐ | ☐ |
| Lighting Director | | ☐ | ☐ |
| | | ☐ | ☐ |
| Audio Director | | ☐ | ☐ |
| Microphone Boom Operator | | ☐ | ☐ |
| Microphone Grip | | ☐ | ☐ |
| | | ☐ | ☐ |
| | | ☐ | ☐ |
| Floor Director | | ☐ | ☐ |
| | | ☐ | ☐ |
| Teleprompter Operator | | ☐ | ☐ |
| | | ☐ | ☐ |
| Camera Operator 1 | | ☐ | ☐ |
| Camera Operator 2 | | ☐ | ☐ |
| Camera Operator 3 | | ☐ | ☐ |
| Camera Operator 4 | | ☐ | ☐ |
| | | ☐ | ☐ |
| Telecine Operator | | ☐ | ☐ |
| | | ☐ | ☐ |
| Videotape Recorder Operator | | ☐ | ☐ |
| | | ☐ | ☐ |
| Video Engineer | | ☐ | ☐ |
| | | ☐ | ☐ |
| | | ☐ | ☐ |
| | | ☐ | ☐ |

# VIDEO SCRIPT

## MULTIPLE CAMERA VIDEO COMMERCIAL PRODUCTION

Producer:

Director:

Commercial Title:
Client:
Length:     :
Date:     /   /
Page     of

| VIDEO | AUDIO |
| --- | --- |
| | |

| VIDEO | AUDIO |
|-------|-------|
|       |       |

# STORYBOARD

MULTIPLE CAMERA VIDEO COMMERCIAL PRODUCTION

Producer:

Director:

Commercial Title:
Client:
Length:       :
Date:       /     /
Page       of

# VIDEO SCRIPT/STORYBOARD
## MULTIPLE CAMERA VIDEO COMMERCIAL PRODUCTION

Producer:

Director:

Commercial Title:
Client:
Length:       :
Date:      /    /
Page      of

| VIDEO | AUDIO |
|---|---|

1
2
3
4
5
6
7
8
9
10
11
12
13
14
15
16
17
18
19
20
21
22
23
24
25
26
27
28
29
30
31
32
33
34
35

| VIDEO | AUDIO |
| --- | --- |

36
37
38
39
40
41
42
43
44
45
46
47
48
49
50
51
52
53
54
55
56
57
58
59
60
61
62
63
64
65
66
67
68
69
70
71
72
73
74
75
76

68

# CHARACTER GENERATOR COPY
## MULTIPLE CAMERA VIDEO COMMERCIAL PRODUCTION

Production Assistant:      Date:   /   /    Page    of

**ACADEMY
LEADER**

Title: [_____]
Length:     :
Client: [_____]
Agency: [_____]
Date:   /   /

Producer: [_____]
Director: [_____]

Record No.:

Font Style:
Font Size:
Color:
Effects:
Record No.:
Front Time:     :
Dialogue Cue:

ORDER [____]
Rolled ☐    Speed ☐
Advanced ☐ Time    :

Font Style:
Font Size:
Color:
Effects:
Record No.:
Front Time:     :
Dialogue Cue:

ORDER [____]
Rolled ☐    Speed ☐
Advanced ☐ Time    :

Font Style:
Font Size:
Color:
Effects:
Record No.:
Front Time:     :
Dialogue Cue:

ORDER [____]
Rolled ☐    Speed ☐
Advanced ☐ Time    :

Font Style:
Font Size:
Color:
Effects:
Record No.:
Front Time:     :
Dialogue Cue:

ORDER ☐
Rolled ☐    Speed ☐
Advanced ☐ Time    :

Font Style:
Font Size:
Color:
Effects:
Record No.:
Front Time:     :
Dialogue Cue:

ORDER ☐
Rolled ☐    Speed ☐
Advanced ☐ Time    :

Font Style:
Font Size:
Color:
Effects:
Record No.:
Front Time:     :
Dialogue Cue:

ORDER ☐
Rolled ☐    Speed ☐
Advanced ☐ Time    :

Font Style:
Font Size:
Color:
Effects:
Record No.:
Front Time:     :
Dialogue Cue:

ORDER ☐
Rolled ☐    Speed ☐
Advanced ☐ Time    :

# GRAPHIC DESIGN

MULTIPLE CAMERA VIDEO COMMERCIAL PRODUCTION

Producer:
Director:
Date:   /   /
Graphic Artist:

Commercial Title:
Client:
Deadline:   /   /

Graphic No. ☐

Title/Label:
Concept:

Text:

Image Design:

Graphic No. ☐

Title/Label:
Concept:

Text:

Image Design:

Graphic No. ☐

Title/Label:
Concept:

Text:

Image Design:

Graphic No. ☐

Title/Label:
Concept:

Text:

Image Design:

Graphic No. ☐

Title/Label:
Concept:

Text:

Image Design:

Graphic No. ☐

Title/Label:
Concept:

Text:

Image Design:

Graphic No. ☐

Title/Label:
Concept:

Text:

Image Design:

# SCRIPT BREAKDOWN
MULTIPLE CAMERA VIDEO COMMERCIAL PRODUCTION

Producer:

Director:

Date:     /     /

Commercial Title:
Client:
Length:        :
Script Length:          pages
Script Lines:           lines
Page            of

| SCRIPT PAGES/ LINES | INT/ EXT | TIME | SET | PROPERTIES | CAST | SHOOT- ING ORDER |
|---|---|---|---|---|---|---|
| | | | | | | |
| | | | | | | |
| | | | | | | |
| | | | | | | |
| | | | | | | |
| | | | | | | |
| | | | | | | |
| | | | | | | |
| | | | | | | |
| | | | | | | |
| | | | | | | |
| | | | | | | |
| | | | | | | |
| | | | | | | |
| | | | | | | |
| | | | | | | |
| | | | | | | |
| | | | | | | |
| | | | | | | |
| | | | | | | |
| | | | | | | |
| | | | | | | |
| | | | | | | |
| | | | | | | |
| | | | | | | |
| | | | | | | |
| | | | | | | |
| | | | | | | |
| | | | | | | |
| | | | | | | |
| | | | | | | |
| | | | | | | |

# TALENT AUDITION

| Commercial Title: | Client: |
|---|---|

Actor/Actress:                             Telephone:
Address:                                        Home ( )    -
City:                                              Work ( )    -
State:                    Zip Code:          Birthdate:     /    /

Availability                                Unavailability
  Dates:                    Hours:              Dates:                         Hours:

Personal Data:  Height              Weight                 Ethnicity:
  Hat Size:           Shoe Size:           Waist Measurement:          Chest Measurement:
  In Seam:            Shirt/Blouse Size:                Suit/Dress Size:

Agent:
Union Affiliation:

Please explain your interest in this production:

Please list the roles/parts for which you feel qualified/interested:

Are you willing to perform as an extra for this production:

Are there other production tasks/roles in which you might be interested (e.g., costumes, make-up, properties, etc.):

PLEASE ATTACH YOUR RESUME AND PHOTO TO THIS FORM, OR LIST PREVIOUS EXPERIENCE
AND/OR DRAMA TRAINING ON THE REVERSE SIDE OF THIS FORM.

# CHARACTERIZATION

MULTIPLE CAMERA VIDEO COMMERCIAL PRODUCTION

Director:                          Commercial Title:
                                   Client:
Actor/Actress:                     Date:      /     /

Character:

List descriptive adjectives:

| Order of importance | Order revealed to audience |
|---|---|
| 1. | 1. |
| 2. | 2. |
| 3. | 3. |
| 4. | 4. |
| 5. | 5. |
| 6. | 6. |
| 7. | 7. |
| 8. | 8. |

List actions performed by the character/done to the character in the commercial:

| Action | Significance |
|---|---|
| 1. | 1. |
| 2. | 2. |
| 3. | 3. |
| 4. | 4. |
| 5. | 5. |

What are the major motivations of the character by script line(s):
    Line(s):
    Line(s):
    Line(s):

What is this character's relationship to all other characters in the commercial?

| Character | Relationship |
|---|---|
| 1. | |
| 2. | |
| 3. | |
| 4. | |
| 5. | |
| 6. | |

## Characterization (Continued)

Label and elaborate on the function the character plays in the commercial:

Describe the audience's emotional reaction to the character at important moments during the commercial:

| The Moment | Audience's Emotional Reaction |
| --- | --- |
| | |

Describe the change in the character from the beginning to the end of the commercial:

How can the change(s) be conveyed to the audience:

Write a birth to death autobiography of the character:

# SET DESIGN

Set Designer:
Producer:                Director:                Commercial Title:
    Approval ☐              Approval ☐          Client:
                                                Date:   /   /          Page        of

SET:                                        SCRIPT LINE(S):

*Bird's Eye View*                                    21 Units x 34 Units

*Front View*                                         12 Units x 34 Units

# COSTUME DESIGN

## MULTIPLE CAMERA VIDEO COMMERCIAL PRODUCTION

| | | |
|---|---|---|
| Costume Master/Mistress: | | Commercial Title: |
| | | Client: |
| Producer: | Director: | Date: / / |
| Approval ☐ | Approval ☐ | Page of |

Character:
Script Unit(s):                    Costume No.:

COSTUME DESCRIPTION                    COSTUME SKETCH

Historical Period:
Costume Type:
Fabric:
Color Scheme:
Trim:
Costume Elements List:
      1.
      2.
      3.
      4.
      5.
      6.
Costume Accessories:
      1.
      2.
      3.
      4.
      5.

NOTES

# MAKE-UP DESIGN

### MULTIPLE CAMERA VIDEO COMMERCIAL PRODUCTION

Make-Up Artist:

Producer:               Director:
  Approval ☐         Approval ☐

Commercial Title:
Client:
Date:      /     /
Page        of

Character:
  Age:          Complexion:
  Sex: M/F       Type:

Actor/Actress:

☐ Wig:
☐ Moustache:
☐ Beard:

| MAKE-UP ELEMENTS | TYPE/CODE | | NOTES |
|---|---|---|---|
| Make-up base | | | |
| Highlighting | | | |
| Shadow | | | |
| Rouge | | | |
| Eye shadow | | | |
| Eye liner | | | |
| Eyebrow pencil | | | |
| Eyelashes | | | |
| Lip rouge | | | |
| Lip liner | | | |
| Highlighting liner | | | |
| Powder | | | |
| Liquid latex | | | |
| Crepe Hair | | | |

Required make-up changes:
  Script line(s):                    Script line(s):
  Make-up changes:                   Make-up changes:

Special requirements:

# PROPERTIES BREAKDOWN
### MULTIPLE CAMERA VIDEO COMMERCIAL PRODUCTION

Property Master/Mistress:

Commercial Title:
Client:

Producer:          Director:

Approval ☐          Approval ☐

Script Length:          pages
Script Lines:          lines
Date :    /    /          Page          of

| SCRIPT PAGE / LINE | CHARACTER(S) | PROPERTIES | NOTES |
|---|---|---|---|
| | | | |
| | | | |
| | | | |
| | | | |
| | | | |
| | | | |
| | | | |
| | | | |
| | | | |
| | | | |
| | | | |
| | | | |
| | | | |
| | | | |
| | | | |
| | | | |
| | | | |
| | | | |
| | | | |
| | | | |
| | | | |
| | | | |
| | | | |
| | | | |
| | | | |
| | | | |
| | | | |
| | | | |
| | | | |
| | | | |
| | | | |
| | | | |
| | | | |
| | | | |
| | | | |
| | | | |
| | | | |

Insurance Required:
  Property:          Type of insurance:
  Property:          Type of insurance:

# BLOCKING PLOT

MULTIPLE CAMERA VIDEO COMMERCIAL PRODUCTION

Director:
Date:  /  /
Set:

Commercial Title:
Client:
Script Page ☐          Script Line ☐

Lighting
☐ Interior
☐ Exterior
☐ Day
☐ Night

Sound
☐ Synchronous
☐ Silent
☐

Bird's Eye Floor Plan

Cameras /Properties/Blocking

Description of Take/Unit          Actor(s)/Actress(es)/Cameras/Movement/Properties

In-Cue                    Dialogue/Action    Out-Cue                    Dialogue/Action

Comments:

| Set: | Script Page ☐ | Script Line ☐ |

Bird's Eye Floor Plan                              Cameras /Properties/Blocking

**Lighting**
☐ Interior
☐ Exterior
☐ Day
☐ Night

**Sound**
☐ Synchronous
☐ Silent
☐

| Description of Take/Unit | Actor(s)/Actress(es)/Cameras/Movement/Properties |

| In-cue | Dialogue/Action | Out-cue | Dialogue/Action |

Comments:

# LIGHTING PLOT

MULTIPLE CAMERA VIDEO COMMERCIAL PRODUCTION

Lighting Director:

Producer:                 Director:

   Approval ☐           Approval ☐

Set:

Commercial Title:

Client:

Script Page ☐        Script Line ☐

Date:    /    /        Page      of

Lighting
☐ Interior
☐ Exterior
☐ Day
☐ Night

Lighting
   Change
☐ Yes
☐ No

Bird's Eye Floor Plan                    Cameras /Properties/Blocking

| Description of Take | Actor(s)/Actress(es) /Cameras/Movement/Properties |
|---|---|
|  |  |

Lighting Instruments          Filters                  Property Lights
  Key Lights:                  Spun Glass:               Lamps:
  Fill Lights:                 Gels:                     Ceiling:
  Soft Lights:                 Scrims:                   Other:

Lighting Accessories          Windows
  Barn Doors:                  Daylight:
  Flags:                       Nighttime:
  Gobo:                        Dusk:
                            Change:

| In-cue | Dialogue/Action Lighting Change Cue | Out-cue | Dialogue/Action Lighting Change Cue |
|---|---|---|---|
|  |  |  |  |

# EFFECTS/MUSIC BREAKDOWN
MULTIPLE CAMERA VIDEO COMMERCIAL PRODUCTION

Audio Director:

Producer:      Director:

Approval ☐     Approval ☐

Commercial Title:

Client:

Date:  /  /      Page    of

| SCRIPT PAGE/LINE | In-cue/Out-cue | Effect | Est. Length | Music |
|---|---|---|---|---|
|  |  |  | : |  |
|  |  |  | : |  |
|  |  |  | : |  |
|  |  |  | : |  |
|  |  |  | : |  |
|  |  |  | : |  |
|  |  |  | : |  |
|  |  |  | : |  |
|  |  |  | : |  |
|  |  |  | : |  |
|  |  |  | : |  |
|  |  |  | : |  |
|  |  |  | : |  |
|  |  |  | : |  |
|  |  |  | : |  |
|  |  |  | : |  |
|  |  |  | : |  |

Rights/Clearance Required:

Effect/Music:        Publisher:        Rights Owner:

Effect/Music:        Publisher:        Rights Owner:

# AUDIO PLOT

## MULTIPLE CAMERA VIDEO COMMERCIAL PRODUCTION

Audio Director:

Producer:                    Director:

    Approval ☐              Approval ☐

Set:

Commercial Title:

Client:

Script Page ☐          Script Line ☐

Date:  /  /              Page      of

### Lighting
☐ Interior
☐ Exterior
☐ Day
☐ Night

### Sound
☐ Synchronous
☐ Silent
☐ _____

### Microphone
☐ Directional
☐ Wireless
☐ _____

### Microphone Support
☐ Fishpole
☐ Giraffe
☐ Handheld
☐ Hanging
☐ _____

### Sound Effects
☐ Foldback
☐ _____

Bird's Eye Floor Plan

Cameras /Properties/Blocking
Microphone Grip(s)/Microphone Support(s)/Cable Run(s)
Sound Perspective: ☐ close       ☐ distant

| Description of Take | Actor(s)/Cameras/Movement/Properties |
|---|---|

| In-cue | Dialogue/Action | Out-cue | Dialogue/Action |
|---|---|---|---|

Rights/Clearance Required:

# AUDIO PICKUP PLOT
MULTIPLE CAMERA VIDEO COMMERCIAL PRODUCTION

Mike Grip 1:

Mike Grip 2:

Audio Director:

    Approval ☐

Commercial Title:

Client:

Date:  /  /

Page   of

SET:

SCRIPT LINE(S):

Bird's eye view of set with properties/actor(s)/actress(es) movement.
Indicate mike grip/audio pickup placement. Placement may change for varying takes/actor movement, etc.

NOTES

# PRODUCTION SCHEDULE

### MULTIPLE CAMERA VIDEO COMMERCIAL PRODUCTION

| | |
|---|---|
| Director: | Commercial Title: |
| Producer: | Client: |
| Production Facility: | Script Length:          pages |
| Date:    /   / | Script Lines:          lines |

PRODUCTION DAY          Date:    /   /          Crew Call:          :          Cast Call:          :

| LINES | STUDIO TIME | SET | CAST | COSTUMES/ PROPERTIES | NOTES |
|---|---|---|---|---|---|
| | : | | | | |
| | : | | | | |
| | : | | | | |
| | : | | | | |
| | : | | | | |
| | : | | | | |
| | : | | | | |
| | : | | | | |
| | : | | | | |
| | : | | | | |

PRODUCTION DAY          Date:    /   /          Crew Call:          :          Cast Call:          :

| | | | | | |
|---|---|---|---|---|---|
| | : | | | | |
| | : | | | | |
| | : | | | | |
| | : | | | | |
| | : | | | | |
| | : | | | | |
| | : | | | | |
| | : | | | | |
| | : | | | | |
| | : | | | | |
| | : | | | | |

# CAMERA SHOT LIST

MULTIPLE CAMERA VIDEO COMMERCIAL PRODUCTION

Camera Operator:
Camera: C1  C2  C3
Studio Production Date:    /    /
Set:

Commercial Title:
Client:
Date:    /    /
Page    of

| SHOT No. | CAMERA FRAMING | CHARACTER/ OBJECT | CAMERA MOVEMENT | SPECIAL INSTRUCTIONS |
|---|---|---|---|---|
| | | | | |
| | | | | |
| | | | | |
| | | | | |
| | | | | |
| | | | | |
| | | | | |
| | | | | |
| | | | | |
| | | | | |
| | | | | |
| | | | | |
| | | | | |
| | | | | |
| | | | | |
| | | | | |

# CONTINUITY NOTES

## MULTIPLE CAMERA VIDEO COMMERCIAL PRODUCTION

Continuity Person:

Taping Date:   /   /

Commercial Title:
Client:
Page      of

| SET | | INTERIOR | DAY | ENDING CAMERA<br>C1    C2    C3 | VIDEOTAPE No. |
|---|---|---|---|---|---|
| | | EXTERIOR | NIGHT | ENDING SHOT No. | SCRIPT LINE No. |
| | | SCRIPT PAGE No. | PRECEDING EDIT LINE No. | | CUT □<br>DISSOLVE □  WIPE □ |
| | DETAILS | | SUCCEEDING EDIT LINE No. | | CUT □<br>DISSOLVE □  WIPE □ |

ACTOR(S)/ACTRESS(ES)/COSTUME/MAKE-UP/PROPERTIES NOTES

| CIRCLE TAKES | 1 | 2 | 3 | 4 | 5 | 6 | 7 | 8 | 9 | 10 | 11 | 12 |
|---|---|---|---|---|---|---|---|---|---|---|---|---|
| END SLATE | | | | | | | | | | | | |
| TIMER | | | | | | | | | | | | |
| COUNTER | | | | | | | | | | | | |
| REASON FOR USE/ NOT GOOD | | | | | | | | | | | | |

ACTION                                    DIALOGUE

# VIDEOTAPE LOG

## MULTIPLE CAMERA VIDEO COMMERCIAL PRODUCTION

Director: _____
Assistant Director: _____
Log Form No. ☐

Commercial Title: _____
Client: _____
Taping Date:   /   /          Page     of

| Script Page | Script Lines | Videotape No. | | Set | | Notes | | | | | | |
|---|---|---|---|---|---|---|---|---|---|---|---|---|
| CIRCLE TAKES | 1 | 2 | 3 | 4 | 5 | 6 | 7 | 8 | 9 | 10 | 11 | 12 |
| END SLATE | | | | | | | | | | | | |
| TIMER | | | | | | | | | | | | |
| COUNTER | | | | | | | | | | | | |
| REASON FOR USE/ NOT GOOD | | | | | | | | | | | | |

| Script Page | Script Lines | Videotape No. | | Set | | Notes | | | | | | |
|---|---|---|---|---|---|---|---|---|---|---|---|---|
| CIRCLE TAKES | 1 | 2 | 3 | 4 | 5 | 6 | 7 | 8 | 9 | 10 | 11 | 12 |
| END SLATE | | | | | | | | | | | | |
| TIMER | | | | | | | | | | | | |
| COUNTER | | | | | | | | | | | | |
| REASON FOR USE/ NOT GOOD | | | | | | | | | | | | |

| Script Page | Script Lines | Videotape No. | | Set | | Notes | | | | | | |
|---|---|---|---|---|---|---|---|---|---|---|---|---|
| CIRCLE TAKES | 1 | 2 | 3 | 4 | 5 | 6 | 7 | 8 | 9 | 10 | 11 | 12 |
| END SLATE | | | | | | | | | | | | |
| TIMER | | | | | | | | | | | | |
| COUNTER | | | | | | | | | | | | |
| REASON FOR USE/ NOT GOOD | | | | | | | | | | | | |

# TALENT RELEASE

## MULTIPLE CAMERA VIDEO COMMERCIAL PRODUCTION

Talent Name: _____
Actor/Actress/Extra          (Please Print)

Commercial Title: _____

For value received and without further consideration, I hereby consent to the use of all photographs, videotapes, or film taken of me and/or recordings made of my voice and/or written extraction, in whole or in part, of such recordings or musical performance

at _____ on _____ 19___
(Recording Location)                    (Month)        (Day)        (Year)

by _____ for _____
(Producer)                          (Producing Organization)

and/or others with its consent, for the purposes of illustration, advertising, or publication in any manner.

Talent Name _____
(Signature)

Address _____ City _____

State _____ Zip Code _____

Date: ___ / ___ / ___

---

If the subject is a minor under the laws of the state where modeling, acting, or performing is done:

Guardian _____ Guardian _____
(Signature)                          (Please Print)

Address _____ City _____

State _____ Zip Code _____

Date: ___ / ___ / ___

# POSTPRODUCTION CUE SHEET
MULTIPLE CAMERA VIDEO COMMERCIAL PRODUCTION

Producer:

Director:

Editor:

Commercial Title:

Client:                    Length:    :

Date:    /    /

| SCRIPT LINES | LOG FORM | TAPE NO. | TAKE NO. | TIME/ CODE | LENGTH | COMMENTS | |
|---|---|---|---|---|---|---|---|
| | | | | : | : | | |
| | | | | : | : | | |
| | | | | : | : | | |
| | | | | : | : | | |
| | | | | : | : | | |
| | | | | : | : | | |
| | | | | : | : | | |
| | | | | : | : | | |
| | | | | : | : | | |
| | | | | : | : | | |
| | | | | : | : | | |
| | | | | : | : | | |
| | | | | : | : | | |
| | | | | : | : | | |
| | | | | : | : | | |
| | | | | : | : | | |
| | | | | : | : | | |
| | | | | : | : | | |
| | | | | : | : | | |
| | | | | : | : | | |
| | | | | : | : | | |
| | | | | : | : | | |
| | | | | : | : | | |
| | | | | : | : | | |
| | | | | : | : | | |
| | | | | : | : | | |
| | | | | : | : | | |
| | | | | : | : | | |
| | | | | : | : | | |
| | | | | : | : | | |
| | | | | : | : | | |
| | | | | : | : | | |
| | | | | : | : | | |

# EFFECTS/MUSIC CUE SHEET
### MULTIPLE CAMERA VIDEO COMMERCIAL PRODUCTION

Audio Director:

Director/Editor:

Commercial Title:
Client:
Date:    /    /
Page        of

| SCRIPT LINE(S) | EFFECT | MUSIC | SOURCE | | LENGTH | IN:CUE / OUT:CUE |
|---|---|---|---|---|---|---|
| | | | | | | IN:CUE |
| | | | | | | OUT:CUE |
| | | | CUT☐ | ☐CUT | | IN:CUE |
| | | | FADE☐ | ☐FADE | : | OUT:CUE |
| | | | CUT☐ | ☐CUT | | IN:CUE |
| | | | FADE☐ | ☐FADE | : | OUT:CUE |
| | | | CUT☐ | ☐CUT | | IN:CUE |
| | | | FADE☐ | ☐FADE | : | OUT:CUE |
| | | | CUT☐ | ☐CUT | | IN:CUE |
| | | | FADE☐ | ☐FADE | : | OUT:CUE |
| | | | CUT☐ | ☐CUT | | IN:CUE |
| | | | FADE☐ | ☐FADE | : | OUT:CUE |
| | | | CUT☐ | ☐CUT | | IN:CUE |
| | | | FADE☐ | ☐FADE | : | OUT:CUE |
| | | | CUT☐ | ☐CUT | | IN:CUE |
| | | | FADE☐ | ☐FADE | : | OUT:CUE |
| | | | CUT☐ | ☐CUT | | IN:CUE |
| | | | FADE☐ | ☐FADE | : | OUT:CUE |
| | | | CUT☐ | ☐CUT | | IN:CUE |
| | | | FADE☐ | ☐FADE | : | OUT:CUE |
| | | | CUT☐ | ☐CUT | | IN:CUE |
| | | | FADE☐ | ☐FADE | : | OUT:CUE |
| | | | CUT☐ | ☐CUT | | IN:CUE |
| | | | FADE☐ | ☐FADE | : | OUT:CUE |
| | | | CUT☐ | ☐CUT | | IN:CUE |
| | | | FADE☐ | ☐FADE | : | OUT:CUE |
| | | | CUT☐ | ☐CUT | | IN:CUE |
| | | | FADE☐ | ☐FADE | : | OUT:CUE |
| | | | CUT☐ | ☐CUT | | IN:CUE |
| | | | FADE☐ | ☐FADE | : | OUT:CUE |
| | | | CUT☐ | ☐CUT | | IN:CUE |
| | | | FADE☐ | ☐FADE | : | OUT:CUE |
| | | | CUT☐ | ☐CUT | | IN:CUE |
| | | | FADE☐ | ☐FADE | : | OUT:CUE |
| | | | CUT☐ | ☐CUT | | IN:CUE |
| | | | FADE☐ | ☐FADE | : | OUT:CUE |
| | | | CUT☐ | ☐CUT | | IN:CUE |
| | | | FADE☐ | ☐FADE | : | OUT:CUE |

# Chapter 4

# Description and Glossary for Commercial Production Organizing Forms

## INTRODUCTION

A barometer to successful television production is the degree to which producing and production tasks are accomplished and the extent to which producing and production tasks are *successfully* accomplished. Equally important to successful accomplishment of those tasks is the knowledge and awareness of the details necessary at each producing and production stage.

This chapter presents a description and glossary of terms for successful studio commercial production task accomplishment. These descriptions and glossaries are presented in terms of the organizing forms found in Chapter 3. Each organizing form is presented here by title, the particular production process to which it refers, the producing or production personnel responsible for the completion of the form and the tasks, and a description of the purpose of the form and of the objective in using the form.

Not all organizing forms are necessary or even required in every television studio commercial production. They are selective and designed to be used as needed. Those forms that organize and assist in accomplishing a studio commercial preproduction, production, or postproduction task should be tried for the ease and thoroughness with which they organize a task. Those forms that are redundant to a particular commercial production should be ignored. The forms are meant to be helpful, not a stumbling block, to studio commercial production.

## DESCRIPTION AND GLOSSARY

### • Marketing Analysis Form

*Production process*   Video commercial preproduction

*Responsibility*   Producer

*Purpose*   To structure the conceptual design of a commercial by analyzing the goals, objectives, and target audience of a proposed commercial and to write a production statement.

*Objective*   The marketing analysis form is designed to encourage a producer to do an analysis of the potential market for the proposed commercial. The proposed market is a function of the goals and objectives of the proposed commercial and of the characteristics of the target audience for the product or service of the commercial. The analysis of the goals, objectives, and target audience should direct the producer to a narrow production statement to be a part of all stages of the commercial's production.

### *Glossary*

**Producer**   The name of the producer of the proposed commercial should be entered here.

**Client Representative**   The name of the person who is the representative of the client or agency whose product or service is to be featured in the proposed commercial is entered here.

**Date**   This is the completion date of the marketing analysis form.

**Commercial Title**   The proposed commercial should have a working title by which it will be known throughout the production process.

**Client**   The name of the client—the manufacturer or company owning the product or service—should be placed here.

**Product/Service**   The name of the product or service being advertised is entered here.

**Goal(s) and Objective(s) for This Commercial**   The goals and objectives a client or client representative has for the product or service being advertised should be listed here. The analysis will be facilitated if every goal or objective is listed regardless of how insignificant it might seem (e.g., simply to make a product or service known to the public is a goal that should be listed).

**Target Audience for This Commercial**  In this space, the producer lists all of the characteristics—demographics—of the audience for whom the product or service is intended. Once demographics are exhausted, the producer should suggest those production values (i.e., those elements of television medium both visual and auditory) that are attractive to the target audience. For example, if the target audience's age demographic is 13 to 18 years, then current teenage music, fashion, colors, and favorite television program elements become production values that may translate eventually into production values for the proposed commercial.

**Production Statement**  The production statement is a succinct, one sentence statement that captures the essence of the proposed commercial. The statement should be so precise that it can be printed on the script, on the video slate, and on any other paperwork for the production as a constant reminder of what the commercial is designed to be and to do.

## • Production Budget Form

*Production process*  Video commercial preproduction

*Responsibility*  Producer

*Purpose*  To realistically estimate all possible costs of the proposed television commercial from preproduction to postproduction.

*Objective*  The budget form is a blank model of a budget for the production of a studio commercial. The form is meant to organize as many facets of multiple camera commercial production as can be anticipated into estimated expenses and, after production, into actual expenses. The form suggests possible cost items across the spectrum of studio video production, personnel, equipment, labor, time, and materials.

The form should be used to suggest expenses and to call attention to possible hidden costs before production begins.

*General comments on the use of the budget form*
The form should be studied for line items that might pertain to the proposed television commercial production. Only applicable line items need to be considered. Use the suggestion of line items to consider as many foreseeable real costs as possible for the project.

Note that all sections of the budget are summarized on the front page of the form. The subtotals of costs from the individual sections are brought forward to be listed in the summary.

Some costs are calculated by the number of days employed, the hourly rate of pay, and overtime hours. Other costs are calculated by the number of items or people and the allotted amount of money per day for the item per person. The cost of materials is calculated by the amount of materials times the cost per unit.

Most cost entry columns of this budget form are labeled "Days, Rate, O/T [overtime] Hrs, Total." For those line items that do not involve day and rate, the total column alone should be used.

Projected costs to be applied to the budget can be determined from a number of sources. One source of equipment rental and production costs is the rate card of a local video production facility. Salary estimates and talent fees can be estimated from the going rates of relevant services of equivalent professionals or from their agencies. Cost quotations can be requested in phone calls to providers of services and materials. Travel costs can be obtained by a phone call to a travel agency or an airline company. Preparing a good budget is going to involve a lot of time and research.

*Glossary*

**Producer**  The name of the producer for the commercial is entered here.

**Director**  The name of the director of the commercial is recorded in this space. If a director has not yet been chosen or hired, the space can be left blank.

**Date**  This is the date of the preparation of the production budget.

**Commercial Title**  The working title for the proposed commercial should be entered here.

**Client**  The client or agency handling the product or service being advertised or the manufacturer of the product or service should be listed here.

**Length**  The length of the proposed commercial should be recorded here.

**No. Preproduction Days**  Estimate the number of days it will take to complete the necessary preproduction stages for the commercial. This estimate includes time on all preproduction stages (e.g., budget, production schedule, script breakdown, casting, master script, and shot list) for all crew members involved in the preproduction of the commercial (e.g., producer, director, camera operators, lighting director, videotape recorder operator, audio director, continuity person, and talent).

**No. Studio Production Days/Hours**  Estimate the number of days and hours for which the studio or control room facilities may be needed in the production of the commercial. This time estimate accounts for all in-studio videotaping (e.g., talent on a set) and the use of the control room switcher or character generator for postproduction.

**No. Postproduction Days/Hours**  Estimate the number of days and hours that will be spent in postproduction in completing the video commercial master edit.

**Studio Production Date**  Indicate the date proposed to begin studio production and videotaping of the commercial.

**Completion Date**  Indicate the projected date of completion of the commercial. This date should include all postproduction tasks.

**Summary of Production Costs**  This summary area brings forward the subtotals of costs from the respective sections within the budget form. Note that each item in lines 1 through 19 is found as a section head within the budget form. When each section is completed, the section subtotal is listed in this summary table.

**Contingency**  This term refers to the practice of adding

a percentage to the subtotal of the summary section (i.e., to lines 1 through 19) as a pad against the difference between actual costs and the estimated costs of all line items. The customary contingency percentage is 15%. Multiply the subtotal in line 20 by 0.15 to determine the contingency figure. Enter that result into line 21. The total of lines 20 and 21 becomes the grand total of the budget costs.

**Estimated**   The *estimated* category refers to costs that can only be projected before the production. Rarely are the estimated costs the same as the final costs because of the number of variables that cannot be anticipated. Every attempt should be made to ensure realistic estimated cost figures and the consideration of as many variables as possible. The challenge of preparing an estimated budget is to project as close as possible to the actual costs.

**Actual**   The *actual* category records the costs that are finally incurred for the respective line items during or after the production is completed. The actual costs are those costs that will really be paid. The ideal budget seeks to have actual costs come as close as possible to— i.e., be at or, preferably, under—the estimated costs.

**Rights and Clearances**   This section accounts for the costs that might be incurred in obtaining necessary copyright clearances, synchronization rights, and trademark registration for music, production logo, or script.

**Script**   This section suggests some of the possible costs that might be incurred with the writing or rewriting of the script for the commercial. On some story lines, searches may have to be made for legal purposes (e.g., product or service claims or libel).

**Producer and Staff**   This section accounts for the producer and other personnel that may be involved in producing responsibilities. The role of producer is divided into the three areas of a production: preproduction, production, and postproduction.

**Director and Staff**   This section covers the director and personnel associated with the directing responsibilities. As with most budgets, some line items (e.g., secretary) are listed for purposes of suggesting enlarged staffs and additional roles that might be needed during multiple camera commercial production. Not all of the line items are necessary to all studio productions.

**Talent**   The expenses of any talent are accounted for in this section. Some talent may be hired by contract and may be subject to other fees (e.g., Screen Actors Guild (SAG)). Other talent may be freelancers. Space in the budget allows for listing the characters from the commercial script and the actor or actress playing the part.

**Benefits**   Depending on the arrangements made with the talent, some production budgets may have to account for benefits accruing to the talent and crew members (e.g., health benefits or pensions). Insurance coverage and state and federal taxes that are due to any salaries to talent and crew members should be accounted for in this section.

**Production Facility**   This section lists production facility space that may be required and, for each facility space, the technical hardware that may or may not be needed. Rate cards for production facilities provide cost infor-

mation for these spaces and for the technical hardware in them. The producer will have to choose what is needed and determine the projected cost of each item.

**Production Staff**   This section accounts for production staff and crew members for each stage of the production: preproduction, production, and postproduction.

**Cameras, Videotape Recording, and Video Engineering**   Those personnel and expenses associated with the maintenance and use of the video camera are accounted for in this section. This section records the equipment that will need to be purchased or rented. Camera equipment rental will be necessary if any exterior stock shots have to be made.

**Set Design, Construction, and Decoration**   Studio commercial production will involve the use of studio sets of some sort (even the cyclorama is considered a set). Existing studio sites or constructed sets with artwork require that some consideration be made for set design, construction, and decoration. This section of the budget suggests some design and construction personnel and some construction and decoration costs.

**Properties**   Many studio commercials require set properties, action properties (e.g., vehicles or animals), or hand properties (e.g., preparation of food). This section of the budget suggests and accounts for personnel and action properties, their care and handling, and the same for hand properties.

**Set Lighting**   The need to create mood and setting on sets requires personnel and equipment for set lighting. Studio costs can be incurred with the use of light instruments. The light instruments necessary to light the set must be listed in the budget.

**Audio Production**   Audio production and recording are essential to commercial production. This section suggests personnel, hardware, microphones, and the need for equipment rental and purchase line items for the production.

**Costuming**   Commercial production requires costumes. The personnel, materials, production, and care of costumes are accounted for in this section.

**Make-up and Hairstyling**   Make-up and hair preparation are necessary to commercial production. This section of the budget suggests line items to be accounted for when make-up, hairstyling, and supplies are to be used.

**Videotape and Film Stock, Film Processing**   The provision of necessary videotape stock in required sizes, film, and film processing is accounted for in this section. The producer has to consider still photography surrounding the commercial and the product as part of production.

**Postproduction Editing**   The personnel, facilities, and materials for postproduction needs during editing are accounted for in this section.

**Music**   Most commercial production will involve original or recorded music. This section of the budget suggests some of the potential costs and personnel involved in the music.

**Artwork and Photography**   It is not uncommon that artwork will be part of the production of any commercial. Artwork means anything from reconstructing the packaging of a product for videotaping purposes to cel animation. The need for still photographs of the studio set and talent for print advertising and still shots of the advertised product are suggested here.

• **Facility Request Form**

*Production process*   Video commercial preproduction

*Responsibility*   Producer

*Purpose*   To organize and request television production space and technical hardware from a production facility.

*Objective*   The facility request form allows the producer to choose from among all possible available television production facilities and television production hardware for the commercial's production. The form prompts the producer regarding facility space and equipment, preproduction requirements, and production crew.

*Glossary*

**Production Facility**   The name of the production facility at which the commercial will be videotaped is entered here.

**Producer**   The name of the producer of the commercial is noted here.

**Director**   The name of the director of the commercial is entered here.

**Date**   The date on which the facility request form was created is logged here.

**Commercial Title**   The title of the commercial is listed here.

**Client**   The client or agency for whom the commercial is being produced is recorded here.

**Length**   The broadcast length is indicated here.

**Production Dates**   The range of dates if production will take more than one day is entered here.

**Facility Space**   The producer chooses from among the possible facility spaces those that are required.

**Studio Requirements**   The producer checks those elements that will be required in the production of the commercial. In choosing the light instruments needed, the producer should indicate the number of such instruments needed and the required kilowatts of the instruments.

**Control Room Requirements**   The producer chooses the equipment in the control room that will be needed.

**Master Control Area Requirements**   The equipment needed from master control is checked here. If more than one unit is required, the number of units should be entered instead of a check mark.

**Audio Control Room Requirements**   The producer must check the equipment from available audio control equipment that will be required for the commercial. As with other choices, if more than one unit is required, the number of units should be entered in the box instead of a check mark.

**Preproduction Requirements**   The producer must indicate to the production facility what preproduction time will be required. Preproduction requirements include time for set construction, set lighting, and rehearsals.

**Studio Production Personnel**   The producer must check, from the production facility management, which production crew personnel will be supplied by the facility and which will have to be supplied by the producing staff. This area of the form permits listing the names of the crew members and checking from which source they are coming.

• **Video Script Form**

*Production process*   Video commercial preproduction

*Responsibility*   Producer

*Purpose*   To coordinate in side-by-side columns audio copy and a verbal description of the video imaging of the proposed commercial.

*Objective*   The preproduction script is the proposed audio copy and the verbal description of the proposed commercial. The audio copy is an important part of the production of the commercial. Rational and emotional appeals and claims have to be weighed carefully. The audio copy is usually written first. In the simple two-column television script, a verbal description of the proposed video is adequate at this stage.

*Glossary*

**Producer**   The name of the producer of the proposed commercial is entered here.

**Director**   When a director is chosen, the director's name is listed here.

**Commercial Title**   Commercials are usually titled and the title is used throughout the production.

**Client**   Since commercials are usually marketing the product or services of some client, the client for the proposed commercial is listed here.

**Length**   Commercials are distinguished by their length. The length of the proposed commercial should be listed here.

**Date**   The date of creation of this form should be entered here.

**Page   of**   This records the expected number of pages of script and the number of each individual page.

**Video**   The video column should contain an abbreviated verbal description of the video elements of the commercial. This column should contain image content descriptions and framing (e.g., XLS of New York City Skyline), editing transitions (e.g., dissolve, cut, or fade), and special effects (e.g., DVE rotation), cut-ins, cutaways, and character generator copy. For example:

VIDEO
XLS New York City Skyline;
Diss to MS house front;
Cut to CU Mrs. Betty Kasper
Super LT          MRS. BETTY KASPER
                          Artist

**Audio**   The audio column should contain the full verbal copy to accompany the video described in the video column. Audio copy is written first in the development of a script. Audio copy should be typed in all caps. All talent and production directions should be typed in upper and lower case. All audio copy should be introduced with the label of the talent intended to deliver the copy. The audio column should also record the use of sound effects, music, and ambience. For example:

AUDIO
SFX: City street traffic sounds
ANN: NOT EVERYONE LIVING IN NEW YORK CITY SUFFERS FROM IMPERSONALIZATION OF THE BIG APPLE.

• **Storyboard Form**

*Production process*    Video commercial preproduction

*Responsibility*    Producer

*Purpose*    To assist in the imaging, flow, and pacing of a proposed commercial.

*Objective*    The storyboard form coordinates in aspect ratio form each change of screen image and audio copy in sequential order. The storyboard facilitates the communication of concept and image of a commercial to the client, the director, and the camera operators. A storyboard encourages a producer and a director to make a commitment to video elements of screen content, framing, and picturization.

*Glossary*

**Producer**    The name of the producer of the proposed commercial is entered here.
**Director**    When a director is chosen, the director's name is listed here.
**Commercial Title**    Commercials are usually titled and the title is used throughout production.
**Client**    Since commercials are usually marketing the product or services of some client, the client for the proposed commercial is listed here.
**Length**    Commercials are distinguished by their length. The length of the proposed commercial should be listed here.
**Date**    The date of creation of this form should be entered here.
**Page of**    This records the expected number of pages of storyboards and the number of each individual page.
**Aspect Frames**    Each aspect ratio frame should be sketched with the basic form or design of the proposed content of the commercial. The basic form or design should approximate the desired video framing expected by the camera. A new frame should be sketched for every significant image change.

Transitions between frames other than the cut, can be indicated in an in-between frame.

Frames should be numbered consecutively in the boxes provided for that purpose at the upper left-hand corner of each aspect ratio frame for easy reference. The corresponding audio copy should be recorded in the boxes provided for that purpose below each frame.

• **Video Script/Storyboard Form**

*Production process*    Video commercial preproduction

*Responsibility*    Producer

*Purpose*    To coordinate a verbal description of the proposed commercial and the proposed audio copy with the corresponding storyboard frames.

*Objective*    The video script and storyboard combination form is an alternative script form in which each storyboard frame is coordinated with its corresponding verbal description and audio copy.

*Glossary*

**Producer**    The name of the producer of the proposed commercial is entered here.
**Director**    When a director is chosen, the director's name is listed here.
**Commercial Title**    Commercials are usually titled and the title is used throughout production.
**Client**    Since commercials are usually marketing the product or services of some client, the client for the proposed commercial is listed here.
**Length**    Commercials are distinguished by their length. The length of the proposed commercial should be listed here.
**Date**    The date of creation of the form should be entered here.
**Page of**    This records the expected number of pages of script and the number of each individual page.
**Video**    This column is used in a manner similar to the video column in the two-column script form. It should contain a simplified verbal description of the video content, camera framing, edit transition, and character generator copy of the proposed commercial.
**Storyboard Frame**    Those storyboard frames should be used that coordinate with the entry in the video column. Not all frames will be used or needed. Simply skip those frames that do not match the video and audio entries.

See the directions for using the storyboard aspect ratio frames in the glossary for the storyboard form.
**Audio**    This column is used in a manner similar to the audio column in the two-column script form. It will contain all audio copy for the proposed commercial. As with any television script, audio copy should be written before the video descriptions opposite the audio cue are created. On this form the storyboard frame should be sketched corresponding to the first line of each new video column entry.
**Line Enumeration**    The lines in the audio column are numbered consecutively. This is done to assist the director in deciding script production units. A production unit in commercials is defined by the number of lines of audio or sound effects copy. These numbered lines are used to make that designation.

• **Character Generator Copy Form**

*Production process*    Video commercial production and postproduction

**Responsibility**   Producer, director, production assistant

**Purpose**   To explicitly define and describe all of the character generator copy to be used for the commercial, including slate, title, graphics, and screen texts.

**Objective**   The character generator copy form organizes and permits the design and control of all video screen text. This form is very specific and detailed in an attempt to be accurate and thorough in the design and production of all character generator copy.

**Glossary**

**Production Assistant**   The production assistant's name is entered here.

**Date**   The date of completion of the character generator copy form is entered here.

**Page of**   This notation indicates the expected number of pages of character generator copy and the number of each individual page. This preserves the order of text over numerous pages of copy.

**Slate**   This model aspect ratio frame labels the important data to be contained on the character generator slate to be used before every videotaped take during production. This information is the least amount of copy necessary. Some production operations may require additional copy. The production assistant should check with the producer or director on the proper data for each production.

**Record No.**   This entry permits a record of the assigned number for entered data pages in a character generator.

**Product Name**   The name of the product or service being advertised is the first titling screen copy.

**Order**   This entry permits a change of order of presentation of screen text information. In these boxes, the desired sequential order of presentation should be noted.

**Rolled/Speed**   These entries make the choice to roll the screen text pages up the screen and indicate the speed chosen at the character generator for the roll.

**Advanced/Time**   These entries make the choice other than rolling title and credits. This entry indicates the choice to advance one screen text page at a time (i.e., a cut). The time indicated is the time allowed for each screen page to be on screen.

**Font Style**   This notation permits the production assistant to design or select a particular font style for the screen text that may be available on the character generator.

**Font Size**   The choice of font size is also available on the character generator and must be decided upon.

**Color**   Character generators also permit the colorizing of text fonts. That choice is indicated here.

**Effects**   Font style effects can also be chosen (e.g., border or drop shadow). Some of these effects may have to be generated at the switcher. The technical director must be alerted to those choices.

**Record No.**   This entry records the page number assigned to the screen text entered for that page.

**Front Time**   This notation records the time for this particular screen to be matted onto the screen measured from the front of the production (i.e., time from first video).

**Dialogue Cue**   Some screen texts will be matted by dialogue or music cues. Those cues should be noted here.

*Note:* Graphics boxes designed in these video screen frames are chosen to indicate a suggested position for the corresponding video text. They are only suggested positionings. If they are chosen to be used, the text could be lettered in the boxes themselves. Otherwise, multiple generic frames are included for creative design and positioning by the production assistant. Since text accuracy is demanded when, for example, personal names are being matted on the screen, it is highly recommended that the character generator text be *typed* onto the form.

## • Graphic Design Form

**Production process**   Video commercial preproduction

**Responsibility**   Producer

**Purpose**   To design those graphics needed as a matte or full screen video source during production.

**Objective**   The graphic design request form allows the producer to request graphics for the proposed commercial. Graphics or graphic design may be required for the title, the product or service being advertised, a mailing address, or a telephone number. These graphics are designed before final client approval.

**Glossary**

**Producer**   The name of the producer proposing the commercial is entered here.

**Director**   The name of the director of the commercial production is entered here.

**Date**   The date of the design of the title and opening is entered here.

**Graphic Artist**   The name of the graphic artist responsible for the creation of the graphics is entered here. The graphic artist may be a freelancer or an artist from an outside art company.

**Commercial Title**   The title of the proposed commercial is entered here.

**Page of**   The expected number of pages for the graphic design form is indicated here as well as each individual page number.

**Deadline**   The producer sets a deadline for completion of the requested graphics.

**Graphic No.**   The graphics being requested can be numbered consecutively for purposes of calculating the total number of graphics being requested or for noting the order in which each graphic is to be inserted into the commercial.

**Title/Label**   The producer indicates a title or label for each graphic (e.g., per inquiry screen or telephone number).

**Concept**   By verbalizing the idea of some graphics, the producer can convey how the proposed graphic should be designed.

**Text**   This area of the form permits the producer to indicate any text required on the graphic. The required text should be typed in this area in the exact form in which the text should appear on the graphic.

**Image Design**   This area affords the producer the opportunity to describe the content of a graphic's image or to provide a verbal description of a graphic to be created.

- **Script Breakdown Form**

*Production process*   Video commercial preproduction

*Responsibility*   Director

*Purpose*   To break the script down into script line units, usually according to common sets or cast requirements.

*Objective*   The script breakdown form organizes the commercial script into a studio production shooting order according to differing common criteria for the shoots (e.g., set similarity, cast, and time of day). The script breakdown form will help organize the shooting units by script line criteria for scheduling and managing the studio production.

*Glossary*

**Producer**   The name of the producer of the proposed commercial is entered here.

**Director**   When a director is chosen, the director's name is listed here.

**Date**   The date of creation of this form should be entered here.

**Commercial Title**   Commercials are usually titled and the title is used throughout the production of the commercial.

**Client**   Since commercials are usually marketing the product or services of some client, the client for the proposed commercial is listed here.

**Length**   Commercials are distinguished by their length. The length of the proposed commercial should be listed here.

**Script Length:   pages**   The total number of commercial script pages should be entered here.

**Script Lines:   lines**   The total number of lines of script copy, dialogue, and stage directions included is entered here.

**Page   of**   This records the expected number of pages of script and the number of each individual page.

**Script Pages/Lines**   The number of pages or lines of script copy for a production unit of the script is a determinant of the length of a shooting unit. The page count can be made in whole page and portion of page numbers. The line count should include every line of dialogue and stage directions.

**Int/Ext**   These abbreviations stand for "interior/exterior" and refer to the script demands for an indoor or outdoor setting. Whether the set required is an interior or an exterior set can be a determinant for the script breakdown and studio shooting unit.

**Time**   The time of day or night and the interior/exterior setting requirements should be recorded here.

**Set**   Set means the specific required interior or exterior site or environment for shooting a unit of the commercial (e.g., a set would be the bedroom at the Smiths' house).

**Cast**   This entry records by name the specific cast characters who are required on the set for a particular unit.

**Shooting Order**   When the entire script is broken down into production units, the director can determine the specific units for each production session and determine the unit shooting order for the entire studio production.

- **Talent Audition Form**

*Production process*   Video commercial preproduction

*Responsibility*   Producer, director

*Purpose*   To serve as an application form for prospective actors and actresses who may wish to audition for a part in the cast.

*Objective*   The talent audition form elicits from an auditioning actor or actress all of the applicable information that the producer and director may want to have as they evaluate the videotaped performance of the prospective cast member. The form records personal information from the applicant that will serve such preproduction stages as costume construction and actor/actress availability. The form also requires a resumé and photograph.

*Glossary*

**Commercial Title**   This form requires the commercial designation by title.

**Client**   The name of the client for whom the commercial is being produced should be entered here.

**Actor/Actress**   The prospective cast member places his or her name in this space.

**Address/City/State/Zip Code/Telephone: Home/Work/ Birthdate**   This section collects relevant personal information on the applicant.

**Availability/Unavailability**   This space elicits necessary information on whether the applicant can be available for the preproduction and production dates.

**Personal   Data:   Height/Weight/Ethnicity/Hat   Size/ Shoe Size/Waist Measurement/Chest Measurement/ In Seam/Shirt or Blouse Size/Suit or Dress Size**   This block of information records all possible data on the applicant, which will serve costuming and make-up background should the applicant be cast in the commercial.

**Agent/Union   Affiliation**   This   information   defines professional applicants who are affiliated with a casting agent or an actors' union. Should the producer and director choose a union or agent affiliated cast member, costs may be incurred or expected salary requirements made.

**Questions**   Four self-explanatory questions are asked of the applicant that will assist the producer and director in evaluating the prospective cast member.

**Resumé/Photo/Experience/Training**   This information, which the applicant either attaches (if the applicant has a resumé and photo) or lists (previous experience or training), will help the producer and director judge the applicant.

• **Characterization Form**

*Production process* Video commercial preproduction

*Responsibility* Producer, director, cast

*Purpose* To structure the initial process of developing a character for an actor or actress and the whole cast.

*Objective* The characterization form asks meaningful questions of each cast member, which should serve to define and develop the character required of the actor or actress. By triggering written responses from each cast member, most facets of a character are listed and probed.

*Glossary*

**Director/Actor/Actress** This form may be used by the director in assisting each cast member and by each actor or actress. Their respective names are placed in this space.

**Commercial Title** The title of the commercial is listed here.

**Date** The date on which the characterization form was begun should be placed here.

**Character** The name of the character from the commercial who must be characterized by an actor or actress is listed here.

**Descriptive Adjectives** After reading the script, the director, actor, or actress is asked to list adjectives that describe the character being developed.

**Order of Importance** After listing descriptive adjectives, the adjectives should be listed in the order of importance perceived by the cast member or the director.

**Order Revealed to the Audience** The adjectives should be listed in the order in which the cast member or director perceives them to be revealed to the audience.

**Actions Performed by and Done to the Character** This section requires an analysis of the actions performed by the character and those actions done to the character.

**Action/Significance** The actions performed by and done to the character are listed in the action column; the significance of each action is indicated in the second column.

**Major Motivations of the Character** The major motivations—those things that influence the character—are listed by script unit.

**Character's Relationship to All Other Characters** The relationship of the character to all other characters in the commercial is listed.

**Function the Character Plays in the Commercial** The function the character plays in the commercial is labeled and elaborated here.

**Audience's Emotional Reaction to the Character during the Commercial** The cast member or director describes what the audience's reaction should be to the character at important moments during the commercial.

**Change in the Character during the Commercial** The cast member or director should describe the change that takes place in the character from the beginning to the end of the commercial.

**How to Convey the Changes to the Audience** The cast member or director should indicate how the changes occurring in the character should be conveyed to the audience by the actor or actress.

**Character History** The last step in characterization is to develop a history in autobiographical form of the character before and after the period of the commercial, from birth to death.

• **Set Design Form**

*Production process* Video commercial preproduction

*Responsibility* Set designer

*Purpose* To provide the set designer with a form and uniform grids on which to design studio set(s) from both a bird's eye view and a front view.

*Objective* The set design form provides the set designer the grid on which to prepare the basic sketch of the proposed set(s) for the commercial. The grids are constructed to permit both a bird's eye view (a view from above the set) and a front view (a horizontal view, from the front looking into the set).

*Glossary*

**Set Designer** The name of the set designer is listed here.

**Producer/Director/Approval** The names of the producer and director are entered here. Their respective approvals of the set design(s) are indicated in the given boxes.

**Commercial Title** The title of the commercial is placed here.

**Client** The name of the client for whom the commercial is being produced is entered here.

**Date** The date of the design is entered here.

**Page of** This designation indicates the expected number of set design forms and the number of the individual pages.

**Set** The name or title of the set is indicated here.

**Script Lines** The line units of the script that use the set design are listed here.

**Bird's Eye View** The larger grid is used for the view of the set design from above—the bird's eye view. This view affords the designation of width and depth, angles and corners, and the placement of set properties (e.g., chairs and lamps).

**21 Units × 34 Units** The grid contains 21 units on the vertical and 34 units on the horizontal side. The set designer can assign any convenient uniform unit of measure to the square unit. For example, by assigning two feet per square, the designer has an area of 42 feet by 68 feet with which to work.

**Front View** The smaller grid is used for a frontal view of the proposed set. Placing the frontal view directly below the bird's eye view allows alignment of the set, point for point, from both viewpoints. The frontal view is seen on the horizontal.

**12 Units × 34 Units** This grid contains 12 units on the vertical and 34 units on the horizontal side. The set designer can assign any convenient uniform unit of measure to the square unit.

- **Costume Design Form**

*Production process*  Video commercial preproduction

*Responsibility*  Costume master/mistress

*Purpose*  To provide a standard analysis and description of each costume.

*Objective*  The costume design form records the details of each costume required for the commercial. The form prompts all elements of costume design and measurement that will be needed during rental, purchase of materials, or construction of every costume. The form requires a sketch of the costume.

*Glossary*

**Costume Master/Mistress**  The name of the costume master/mistress is indicated here.
**Producer/Director/Approval**  The names of the producer and director are entered here. Both the producer and director indicate approval of the costume design by signing/initialing their respective approval of the design. The producer's approval authorizes the necessary expenditure to create the costumes.
**Commercial Title**  The title of the commercial is listed here.
**Client**  The name of the client for whom the commercial is being produced is entered here.
**Page of**  This records the total number of costume design forms and the number of each individual page.
**Character**  The character for which a costume is being designed is listed here.
**Script Unit(s)**  The script unit(s) from the commercial script for which the costume is required is listed here.
**Costume No.**  Each costume required for every character is numbered. The number can indicate the successive costume within the commercial for the character. Further coding can serve a cost accounting function for budget purposes (e.g., JD–3 would be the code for John Doe's third costume).
**Costume Description**  The costume description records relevant details of the costume.
**Historical Period**  Historical period places the costume within the period of history of which it is a part.
**Costume Type**  Costume type describes the use of the costume (e.g., evening formal or bed clothes).
**Fabric**  Fabric describes the type of material of which the costume is made.
**Color Scheme**  The color scheme indicates the particular color design of the set, the commercial, and the mood created by the director into which the costume must fit. The matching or contrasting colors that will become part of the costume should be described.
**Trim**  Trim indicates the material chosen as edging, an accessory to the primary material of the costume.

**Costume Elements List**  This list records specific elements to be used as part of the costume or in addition to the costume. This would include such items as shoes, socks, underwear, or suspenders.
**Costume Accessories**  This list records those decorative additions, apart from the essential elements of a costume, such as jewelry, boutonniere, cuff links, or gloves.
**Costume Sketch**  This area invites a simple descriptive sketch of the costume from shoulders to feet.
**Notes**  This block should be used to include any additional elements of the costuming not particularly described previously.

- **Make-up Design Form**

*Production process*  Video commercial preproduction

*Responsibility*  Make-up artist

*Purpose*  To record the individual creative elements of each character's make-up requirements.

*Objective*  The make-up design form affords a listing of all possible make-up design elements that could go into the make-up requirements of each character. The form also records the appearance requirements of the character. The make-up artist can also sketch the make-up design on a generic face form.

*Glossary*

**Make-up Artist**  The make-up artist's name is entered here.
**Producer/Director/Approval**  The names of the producer and director are entered here. The make-up artist requires design approval from both the producer and director. Each signs or initials approval of the design. The producer's approval includes the authorization of expenses for the purchase of necessary make-up supplies.
**Commercial Title**  The title of the commercial is listed here.
**Client**  The name of the client for whom the commercial is being produced is entered here.
**Date**  The date of the make-up design should be noted here.
**Page of**  This notation indicates the expected number of make-up designs and the number of each individual page.
**Character/Actor/Actress**  These entries record the character from the commercial and the actor or actress cast in the role.
**Age/Sex/Complexion/Type**  These categories summarize requirements of the character.
**Wig/Moustache/Beard**  These choices indicate the facial or head hair that may be required.
**Make-up Elements/Type or Code/Notes**  This listing records all possible make-up supplies that a character's make-up design could require. The appropriate box should be checked and the type or code of supply indicated. The notes column permits further elaboration on the make-up supplies, including the amount of an element needed.

**Required Make-up Changes** This box records any changes of make-up that may be required of a character (e.g., aging).

**Script Line(s)/Make-up Changes** The script lines requiring the changes are noted here along with the kind of make-up change required.

**Special Requirements** This area provides space to make any further notations of special make-up requirements.

## • Properties Breakdown Form

*Production process* Video commercial preproduction

*Responsibility* Properties master/mistress

*Purpose* To record from the script all required set, hand, and action properties.

*Objective* The properties breakdown form records by script line unit all of the properties needed for the commercial. Properties can be those elements of furniture to decorate and dress a set, the hand properties that actors and actresses need (e.g., a pen or glass), and action properties (e.g., a bicycle or a dog). The properties master/mistress carefully reads the script to create the breakdown list.

*Glossary*

**Properties Master/Mistress** The name of the properties master/mistress is indicated here.

**Producer/Director/Approval** The names of the producer and director are entered here. The properties master/mistress requires the approval of the properties breakdown list by the producer and director. They indicate their approval by signing or initialing the form. The producer's approval authorizes the acquisition of the approved properties by construction, purchase, or rental.

**Commercial Title** The title of the commercial is listed here.

**Client** The name of the client for whom the commercial is being produced is entered here.

**Script Length: Pages** The length of the commercial in terms of pages is entered here.

**Script Lines: Lines** The length of the commercial in terms of script lines, including both dialogue and stage directions, is entered here.

**Page of** This notation indicates the expected number of pages of the properties breakdown list and the number of each individual page.

**Date** The date of preparation of the properties breakdown list is entered here.

**Script Page/Line** The properties master/mistress indicates each script line unit requiring properties.

**Character(s)** This notation indicates the character who has to use the required property.

**Properties** This column records the required properties.

**Notes** This space allows for further description or labeling of the required properties or their use.

**Insurance Required** This space records notification of the producer by the properties master/mistress that the use of some property requires insurance coverage. The particular property should be listed along with an indication of the type of insurance required.

## • Blocking Plot Form

*Production process* Video commercial preproduction

*Responsibility* Director

*Purpose* To create the director's master script—the studio document from which the actors, dialogue, properties, cameras, camera framing, and composition will be directed.

*Objective* This essential preproduction stage prepares the director for all details for directing the cast and crew while in the studio. From the approved set designs, the director can design all production elements for production. The blocking plot serves to record the blocking of actor(s)/actress(es), cameras, and set(s), and to record the storyboard sketch of every camera frame proposed for production.

*Some notes on the creation of the blocking plot form* Many copies of the blocking plot form may be needed. One bird's eye view blocking frame is contained on each page. Each page of the blocking plot form is designed to be inserted into the director's copy of the master script facing the corresponding page of dialogue.

The director should sketch the bird's eye view of the set from the set design form. The director should then position the actor(s)/actress(es) in the floor plan and indicate their blocking placement and movement. The cameras should be sketched into place to achieve the shot(s) as proposed in the storyboard frames in the right-hand column of the master script.

Each bird's eye floor plan is drawn and inserted adjacent to the master script with the corresponding storyboard frames on the script. The bird's eye floor plan need not be redrawn each time if there are no major changes. Once blocking is indicated on the blocking plot, cameras are added as the storyboard frames are sketched in on the master script pages.

*Glossary*

**Director** The director's name is entered here.

**Date** The date when the blocking plot is done is noted here.

**Set** The name or label of the set for the script units being blocked is entered here.

**Commercial Title** The title of the commercial is listed here.

**Client** The name of the client for whom the commercial is being produced is entered here.

**Script Page** The number of the script page to which the blocking plot refers is recorded here.

**Script Line** The corresponding script line for which the blocking plot is being created is recorded here.

**Lighting** The general lighting cue for the set for this script unit is recorded in this column.

**Interior/Exterior/Day/Night** These are the evident choices for lighting cues to the lighting director.

**Sound** The general sound pickup requirement is recorded here.

**Synchronous/Silent** These possible choices—synchronous if sound is to be picked up with the video on the set and silent if video only—may be required and no sound recorded.

**Bird's Eye Floor Plan** This space is the area in which the bird's eye view of the set is sketched for blocking purposes.

**Cameras/Properties/Blocking** These are the elements to be entered on the bird's eye view of the set. Cameras and actor(s)/actress(es) must be blocked and set properties (e.g., furniture), which are important to the set, must be included on the bird's eye view.

**Description of Take/Unit** This space records any verbal description of the action of the take or script unit. Some blocking may be better described in words than can be shown with sketched blocking.

**Actor(s)/Actress(es)/Cameras/Movement/Properties** These are the elements of the blocked unit that are to be included in any description of the take or unit being blocked.

**In-cue/Dialogue/Action** The in-cue is the dialogue or stage direction action that begins the take or script unit.

**Out-cue/Dialogue/Action** The out-cue is the dialogue or stage direction action on which the take or script unit ends.

**Comments** This section provides more space in which the director can make notes on the blocking for the unit. All thoughts should be jotted down during preproduction lest they be glossed over or forgotten during production.

## • Lighting Plot Form

*Production process*   Video commercial preproduction

*Responsibility*   Lighting director

*Purpose*   To prepare and organize the lighting design of the commercial.

*Objective*   The lighting plot form is designed to facilitate the lighting design for the lighting director for each studio set. The plot helps the lighting director preplan the placement of lighting instruments, kind of lighting, lighting control, and lighting design. When lighting is preplanned, lighting equipment needs are easily realized and provided.

*Notes on the use of the lighting plot form*   The lighting plot form is created to encourage preplanning for lighting design and production. Hence, the more lighting needs and aesthetics that can be anticipated, the better the lighting production tasks. At the minimum, every different studio set lighting design should be created in advance of the production session. Lighting design can begin after the master script is complete. The essentials to planning lighting design are the bird's eye floor plan, the actor(s)/actress(es) blocking, information on whether the set is an interior or exterior setting, the time of day, and the mood of the script.

A different plot is not required for every script line unit videotaped in the same set. Base lighting design need not change as script line units in a set change. However, set changes will require new lighting set-up and design. It is these changes for which the lighting director must prepare.

*Glossary*

**Lighting Director** The name of the lighting director for the production is entered here.

**Producer/Director/Approval** The names of the producer and director are entered here. The lighting plots must be approved by the producer and director. They should sign or initial their approval on this form. Approval by the producer authorizes expenses for lighting needs acquisition.

**Set** The lighting director indicates the name or label for the set being lighted.

**Commercial Title** The title of the commercial is listed here.

**Client** The name of the client for whom the commercial is being produced is entered here.

**Script Page** The lighting director will require an orientation to the respective page(s) of the script for every studio set to be lighted.

**Script Line** The script line unit being produced is another orientation to the script to benefit the lighting director.

**Lighting: Interior/Exterior/Day/Night** This area of the plot notes the basic options for lighting design: inside or outside, day or night.

**Lighting Change: Yes/No** It is important to lighting design and lighting control to know if the script requires or the director plans any change of lighting during the unit being videotaped. This notation alerts the lighting director to that need.

**Bird's Eye Floor Plan** The best preproduction information for the lighting director is the floor plan for the set with actor(s)/actress(es) and camera blocking and movement indicated. This should be sketched in this box from the master script. Where the cameras are to be placed is also important to the lighting director and lighting design.

**Description of Take** Any director's notes on the elements of the take that will affect lighting design should be noted here.

**Lighting Instruments/Lighting Accessories/Filters/Property Lights/Windows** These lists are designed to assist the lighting director in considering all elements of lighting design and materials or situations in preparing the lighting plot. A lighting director can make notations in the proper space in planning for the particular design of each set.

**In-cue** A lighting director may find it convenient to make note of the in-cue from the script when—on what action or dialogue—the lighting design is to begin.

**Out-cue** The same notation on an out-cue of action or dialogue for the end of a shot or a change of lighting may be advantageous to the lighting director.

- **Effects/Music Breakdown Form**

*Production process*   Video commercial preproduction

*Responsibility*   Audio director

*Purpose*   To provide a listing of all of the required sound effects and music for the commercial.

*Objective*   The effects/music breakdown form serves the audio director as the script breakdown form serves the director. The audio director studies the script for all required prerecorded audio tracks, sound effects, and music whether already available or to be creatively designed for the production.

*Glossary*

**Audio Director**   The name of the audio director is entered here.
**Producer/Director/Approval**   The names of the producer and director are entered here. The audio director must have the sound effects and music requirements approved by the producer and director. The approval by the producer authorizes expenses for the acquisition of needed sound effects and music sources.
**Commercial Title**   The title of the commercial is listed here.
**Client**   The name of the client for whom the commercial is being produced is entered here.
**Date**   The date of preparation of the breakdown form is recorded here.
**Page  of**   This notation indicates the expected number of pages to the breakdown and the number of each individual page.
**Script/Page/Line**   The audio director records by either the page or line designation the places in the script where any effect or music is required or designed.
**In-cue/Out-cue**   The audio director records the corresponding in-cue and out-cue from the script dialogue or stage directions for the beginning and the end of any effect or music on the appropriate pages.
**Effect**   This column records the particular sound effect required or designed for the cues noted. Required prerecorded audio track(s) should be noted here.
**Est. Length**   Part of either the required or designed effect or music is the estimated length or duration of the effect or music. That information is recorded here.
**Music**   This column records any required or designed music for the cues noted.
**Rights/Clearances Required**   The audio director indicates what music/effects require copyright clearance and other rights (e.g., synchronization). It is the producer's responsibility to secure the necessary rights.

- **Audio Plot Form**

*Production process*   Video commercial preproduction

*Responsibility*   Audio director

*Purpose*   To facilitate and encourage the design of sound perspective and recording for commercial production.

*Objective*   The audio plot form is designed to prompt the audio director to weight the set environment and sound production values in the planning for audio equipment and quality microphone pickup and recording during production.

***Notes on the use of the audio plot form***   The audio plot form is intended to encourage preproduction by the audio director. One plot for every set (depending on the extent of blocking) is probably adequate, although multiple forms might be required for a very detailed and complicated sound recording take.

*Glossary*

**Audio Director**   The name of the audio director is placed here.
**Producer/Director/Approval**   The names of the producer and director are entered here. The audio director is required to get the approval of the producer and director of the audio plot. The approval of the producer authorizes the expenses to be incurred for audio coverage of the production.
**Commercial Title**   The title of the commercial is listed here.
**Client**   The name of the client for whom the commercial is being produced is entered here.
**Date**   The date the plot was designed is entered here.
**Set**   The set for which the audio plot is designed is listed here.
**Page  of**   This notation indicates the expected number of pages of audio plot forms that have been designed and each individual page number.
**Script Page**   The audio director requires an orientation to the respective page(s) of a script for every set to be covered for sound.
**Script Line**   The script line unit being videotaped is another orientation to the script to benefit the audio director.
**Lighting: Interior/Exterior/Day/Night**   This area of the plot notes the basic options for sound perspective design: inside or outside, day or night.
**Sound: Synchronous/Silent**   This choice of sound recording needs indicates what would be required on the set by the audio director. Synchronous sound indicates that both audio and video will be recorded during a take; silent means that video only is required.
**Microphone: Directional/Wireless/Other**   The audio director can make a choice of microphones to be used during videotaping the proposed script unit.
**Microphone Support: Fishpole/Giraffe/Handheld/Hanging/Other**   The audio director can indicate the proposed method of microphone support for sound coverage during production.
**Sound Effects: Foldback/Other**   Here the audio director can indicate required effects, including foldback to the cast as well as to the microphone grips.
**Bird's Eye Floor Plan**   The best preproduction information for the audio director is the bird's eye floor plan with actor(s)/actress(es) blocking and movement indicated. This should be sketched in this box from the

director's master script. Placement of the cameras is also important information for the audio director in the planning of sound design and microphone and microphone boom grip/operator placement.

**Sound Perspective: Close/Distant**  When a change in framing is indicated it could trigger a change in sound perspective; the change in framing should be calculated into the basic sound design. The new framing composition is indicated by checking the appropriate box for new framing. Quality sound perspective creates the auditory sense that when an actor/actress is framed closely, sound levels should be higher; when an actor/actress is framed at a distance, sound levels should be lower. Close or distant framing will also indicate to an audio director that use of a microphone mount will have to be controlled to avoid catching the mike or mike or boom shadows in the camera shot.

**Description of Take**  Any of the director's notes on the elements of the take that will affect sound design and recording should be noted here. For example, excessive movement of actor(s)/actress(es) or properties and expressive gestures could affect sound recording and control on a set. The fact that some sound playback may be required during a take would be noted here.

**In-cue**  The in-cue, either from the dialogue or action, can affect the sound design of a take. That in-cue should be noted here.

**Out-cue**  The out-cue or final words of dialogue or action can affect sound design. Notation of that out-cue should be made here.

**Rights/Clearance Required**  This space is for the notification of the producer that rights or clearances will have to be obtained for any noted effects or music.

- **Audio Pickup Plot Form**

*Production process*  Video commercial preproduction

*Responsibility*  Microphone boom grip(s)/operator(s)

*Purpose*  To analyze and design proposed sound pickup placement on studio sets before production.

*Objective*  The audio pickup plot form is an attempt to create as much quality preproduction analysis as possible from the production crew to benefit the production of the commercial. With this form, the microphone boom grip(s)/operator(s) can take what preproduction information is available to the grip(s)/operator(s) and plan microphone and operator placement within a set before a production session. By studying the blocking of actor(s)/actress(es) and the placement of cameras, cable runs and operator placement can be proposed before blocking occurs.

*Glossary*

**Microphone Boom Grip/Operator 1 and 2**  These entries list the microphone boom grip(s)/operator(s) assigned to the studio production. It may be that the analysis required for this plot will indicate that more than one operator or grip will be required for the production.

**Audio Director/Approval**  The name of the audio director is entered here. The microphone boom grip(s)/operator(s) will need the audio director's approval for this plot.

**Commercial Title**  The title of the commercial is listed here.

**Client**  The name of the client for whom the commercial is being produced is entered here.

**Date**  The date of the preparation of the plot is indicated here.

**Page  of**  This notation indicates the expected number of plots that are part of the proposed sound coverage and the page number of any one form within the total.

**Set**  The set for which the pickup plot is designed is labeled here.

**Script Line(s)**  The script line unit or units for which the set and pickup plot is proposed are recorded here.

**Bird's Eye View**  The bird's eye view of the set with the set itself, set properties, actor(s)/actress(es), and cameras in place is sketched here. This information is available from the director's master script. The blocking indicated will suggest to the microphone boom grip(s)/operator(s) what areas will have to be covered by the microphone(s) under the control of the operator(s). The microphone boom grip(s) and operator(s) are sketched into the set avoiding the lenses of the cameras.

**Notes**  This area of the audio plot form permits the microphone boom grip(s)/operator(s) to record any other elements or questions on sound coverage before production begins.

- **Production Schedule Form**

*Production process*  Video commercial preproduction

*Responsibility*  Director

*Purpose*  To preplan the script line units to be produced for the production session.

*Objective*  The production schedule form organizes all elements of the commercial for the production session. The director has to plan the number of script line units, set availability, cast availability, and approximate times for the production session in the studio. All producing and production staff will need copies of this schedule in advance of the scheduled production session.

*Glossary*

**Director/Producer**  The names of the director and producer are listed here.

**Production Facility**  The name of the production facility is entered here.

**Date**  The date the production schedule was created is noted here. At this stage of preproduction, changes in the production schedule will be gauged from the later date on the form.

**Commercial Title**  The title of the commercial is listed here.

**Client**  The name of the client for whom the commercial is being produced is entered here.

**Script Length: Pages**  This indication of the length of the script in pages confirms information already held by the cast and crew with scripts.

**Script Lines: Lines** This indication of the number of total lines in the script is also a final confirmation of the extent of the script.

**Production Day** The production schedule form is broken into a number of available production session days.

**Date** The date of the production day is firmed with this entry.

**Crew Call** The director indicates the time of the required rendezvous for the production crew. The crew call may not be at the same time as the cast call. The studio crew, with responsibilities for major production equipment, may need more lead time than does the cast before beginning the first unit blocking.

**Cast Call** The director indicates the time of the required rendezvous of the cast members and staff. The cast call may not necessarily be at the same time as the crew call.

**Lines** This column lists the exact script line(s) that the director plans to videotape during the studio production session.

**Studio Time** In this entry, the director plans the time needed to accomplish each scheduled script unit.

**Set** The director confirms the studio set(s) required for the scheduled script units.

**Cast** The cast that will be required for the script line units scheduled are listed here.

**Costumes/Properties** The director indicates the required costumes and properties (or specialized properties) required for the scheduled script units.

**Notes** The director indicates to the staff, cast, and crew any additional information relevant for the particular production session or script line unit.

- **Camera Shot List Form**

*Production process* Video commercial preproduction

*Responsibility* Director

*Purpose* To translate the proposed storyboard frames from the master script into a shot list for each camera.

*Objective* The shot list form organizes each proposed storyboard frame on the director's master script into shots listed by studio camera. Each shot should generate the necessary video to create the proposed edits as designed on the preproduction script. The shot list form will translate each proposed shot from the master script to individual lists for the camera operators.

*Glossary*

**Camera Operator** The name of the camera operator (each of the three) will be entered on this line.

**Camera: C1, C2, C3** The camera assigned to the particular camera operator is indicated here.

**Studio Production Date** The date of the production session scheduled for the listed camera shots is entered here.

**Set** The set being used for the listed camera shots is recorded here.

**Commercial Title** The title of the commercial is listed here.

**Client** The name of the client for whom the commercial is being produced is entered here.

**Date** The date of the creation of the shot list is indicated here.

**Page of** This records the expected number of pages of shot lists per production session and the number of each individual page.

**Shot No.** Every proposed shot needed to create the storyboard frames of the script should be numbered consecutively on the director's master script. The shot numbers are camera specific. The technical director has to separate the numbered shots from the master script and list them for each camera in the order in which they appeared in the master script. Each camera shot list will contain the numbered shot assigned to each individual camera. Those numbers are listed in this column.

**Camera Framing** Camera framing directions for every proposed shot should make use of the symbols for basic camera shot framing: XLS, LS, MS, CU, and XCU. This will communicate to the camera operator the lens framing for the proposed shot. The framing choice for the shot should be checked on the master script by the technical director. That framing direction should be recorded here.

**Character/Object** The content of each framed shot is indicated here. The master script should contain this information with every storyboard frame sketched. For example, the commercial character Henry might be indicated by the initial "H" under the master script storyboard frame. "Henry," then, should be entered in this column.

**Camera Movement** Camera movement directions indicate the kind of camera movement designed for achieving the proposed shot. Camera movement in a shot can be primary movement. (*Primary movement* is movement on the part of the talent in front of the camera.) This will require camera movement—a pan, tilt, or pedestal—to follow the actor or actress. (This camera movement is called *secondary movement*—the movement of the camera itself. This camera movement also could be a dolly, truck, pedestal, or zoom.) The technical director should translate those directions from the director's master script into this column for the camera operator.

**Special Instructions** The director may indicate some special instructions in the master script or blocking plot for a particular shot or camera operator. Those instructions should be indicated here. The camera shot list is camera specific, so an individual camera operator can be addressed in these special instructions.

- **Continuity Notes Form**

*Production process* Video commercial production

*Responsibility* Continuity person

*Purpose* To record during studio production the details of all elements of the production as an aid to reestablishing details for sequential takes. The need is to facilitate a continuity of production details so editing across video-taped takes is continuous.

*Objective*  The continuity notes form is a record of all possible studio production details (e.g., set, cast, properties, and dialogue) that must carry over from take to take and across edits in postproduction. The nature of production is such that, while videotaped takes are separate in production, the final editing of the takes must look continuous. The continuity notes form records specific details of all facets of production in order to reestablish the detail in subsequent videotaping.

*Noting on the use of the continuity notes form*  The form is designed to be used for each separate take in production. This means that for any single line unit, with many takes, one continuity notes page should be used. In preparing for a lengthy production session involving many line units, many copies of this form will have to be used. The continuity person should be someone who is very observant. The role demands attention to the slightest detail of the production that may have to be reestablished in another studio take. The role requires constant note taking.

Admittedly, the immediacy of replaying video on a studio playback monitor is a fast check on any detail in a previous videotaped take. However, while playback is fast, it is time consuming and is not a habit to which a director should become accustomed. Careful, detailed continuity notes are still a requirement in production.

*Glossary*

**Continuity Person**  The name of the crew member responsible for taking continuity notes is entered here.

**Taping Date**  This notation specifies the production session during which the notes were taken.

**Commercial Title**  The title of the commercial is listed here.

**Client**  The name of the client for whom the commercial is being produced is entered here.

**Page  of**  This entry indicates the expected number of continuity notes forms that were used and the page number of any single form.

**Set**  A description of the set or the environment for the production session is required here. Note should be made of things that were moved (e.g., action properties) in order to be replaced.

**Interior/Exterior**  These choices allow the notation of whether the set is considered indoors or outdoors.

**Day/Night**  These choices allow the notation of time of day of the setting.

**Script Page No.**  For a script unit being produced with numbered pages, the page number of the script page being videotaped is recorded in this space.

**Ending Camera/ C1, C2, C3**  This box quickly records the studio camera with the final shot of the take being noted.

**Ending Shot No.**  This box records the numbered shot taken last.

**Videotape No.**  The continuity person should learn from the videotape recorder operator what videotape stock is being used as the recording source tape and its code number should be recorded here.

**Script Line No.**  The number of the script line being videotaped is recorded here.

**Preceding Edit Line No: Cut/Dissolve/Wipe**  The number of the preceding line edited to the current take is recorded here with the type of visual transition designed for it from the director's master script.

**Succeeding Edit Line No: Cut/Dissolve/Wipe**  Similar to the previous entry, this records the sequential edit unit called for from the master script and the visual transition designed for the edit. This information will indicate the special requirements for continuity details. If a wipe or dissolve is called for, a second B-roll videotape will be required to save a generation of video and a change of videotape code will be required.

**Actor(s)/Actress(es)/Costume/Make-up/Properties  Notes**  This area should be used to verbally record all details noticed during production that may have to be reestablished for a subsequent take or unit of the script. Note should be made of any costume use (e.g., a tie off center or a pocket flap tucked in), make-up detail (e.g., smudged lipstick or position of a wound), or property use (e.g., a half-smoked cigarette in the right hand between the index and middle fingers or a fresh ice cream cone in the left hand).

**Circle Takes (1 through 12)**  This area of the form facilitates recording each subsequent take of a unit. The best use of these boxes is to circle each number representing the take in progress until the take is satisfactorily videotaped.

**End Slate**  The normal practice of indicating on the videotape leader what take of a unit is being recorded is to use the character generator slate. If a particular take is flubbed or messed up for a minor problem (e.g., an actor missing a line), the director might simply indicate that, instead of stopping taping and beginning from scratch, the crew keep going by starting over without stopping for the slate at the beginning of the retake. Since the slate was not recorded at the beginning of the retake, an end slate is used—the slate is recorded on tape at the end of the take.

This note will alert the postproduction crew to look for a slate not at the beginning but at the end of that take.

**Timer/Counter**  This area of the form should be used to record a consecutive stop watch time or the consecutive videotape recorder digital counter number reported by the videotape recorder operator. Both of these times should be a record of the length into the videotape where this take was recorded. Therefore, the stop watch or the counter should be set at zero when the tape is rewound to the beginning.

It is good practice for the videotape recorder operator to call the counter numbers over the intercom to the floor director to inform the continuity person at the beginning of every take or retake.

**Reason for Use/Not Good**  One benefit of the continuity notes is to save the tedium of reviewing all takes after a production session to make a determination of the quality of each take. If detailed notes are made, a judgmental notation on the quality of each take is recorded, and one postproduction chore is complete. Usually, a producer or director makes a judgment anyway in determining whether to retake a unit or to move on to a new unit. The continuity person should note in a word or two whether the take is good or not and the reason for that judgment.

**Action** Continuity details on any action during the take should be noted here. For example, the direction an actor or actress takes when turning or the hand used to open a door should be noted.

**Dialogue** An area of continuity interest can be the dialogue. Notation should be made of any quirk of dialogue used in a previous take that may have to be repeated in a subsequent take intended for a matched edit.

• **Videotape Log Form**

*Production process* Video commercial production

*Responsibility* Assistant director

*Purpose* To record from the control room the decisions across the takes and retakes of scheduled script line units during production.

*Objective* The videotape log is designed to record and note quickly the sequential videotaped takes of each script line unit during production. This log becomes a valuable resource to the producer and director in postproduction by indicating the order of takes on a source videotape as well as the designation of the quality of each take. This form should save the need for reviewing in detail every take recorded on the source videotape.

*Glossary*

**Director** This line lists the director of the commercial.

**Assistant Director** This entry contains the name of the assistant director, who is responsible for logging the videotape information.

**Log Form No.** The videotape log forms should be numbered consecutively for ease in reordering the forms before postproduction.

**Commercial Title** The title of the commercial is listed here.

**Client** The name of the client for whom the commercial is being produced is entered here.

**Taping Date** This notation records the date of the production session.

**Page of** This notation indicates the expected number of log forms for the production and the number of each individual log form.

**Script Page** The assistant director records the page number of the script for the unit being recorded.

**Script Lines** This box records the script lines being videotaped.

**Videotape No.** This entry cross checks the videotape stock being recorded by the videotape recorder operator and assists in organizing resources for postproduction.

**Set** This indicates the name or label of the set being used for the current take.

**Notes** This space permits any additional notations relevant to the takes being videotaped during production session.

**Circle Takes (1 through 12)** This area of the form facilitates recording each subsequent take of a unit. The best use of these boxes is to circle each number representing the take in progress until the take is satisfactorily videotaped.

**End Slate** The normal practice of indicating on the videotape leader what take of a unit is being recorded is to use the character generator slate. If a particular take is flubbed or messed up for a minor problem (e.g., an actor missing a line), the director might simply indicate that, instead of stopping taping and beginning from scratch, the crew keep going by starting over without stopping for the slate at the beginning of the retake. Since the slate was not recorded at the beginning of the retake, an end slate is used—the slate is recorded on tape at the end of the take.

This note will alert the postproduction crew to look for a slate not at the beginning but at the end of that take.

**Timer/Counter** This area of the form should be used to record a consecutive stop watch time or the consecutive videotape recorder digital counter number reported by the videotape recorder operator. Both of these times should be a record of the length into the videotape where this take was recorded. Therefore, the stop watch or the counter should be set at zero when the tape is rewound to the beginning.

It is good practice for the videotape recorder operator to call the counter numbers over the intercom to the assistant director at the beginning of every take or retake.

**Reason for Use/Not Good** One benefit of the videotape log is to save the tedium of reviewing all takes after a studio production session to make a determination of the quality of each take. If detailed log notes are made, a judgmental notation on the quality of each take is recorded and one postproduction chore is complete. Usually, a producer or director makes a judgment anyway in determining whether to retake a unit or to move on to a new unit. The assistant director should note in a word or two whether the take is good or not and the reason for that judgment.

• **Talent Release Form**

*Production process* Video commercial production

*Responsibility* Producer

*Purpose* To give the producer legal rights over the video and audio recording of the talent.

*Objective* The talent release form is a legal document that, when filled out and signed by the talent, gives to the producer and the producing organization the legal right to use both the video and audio recording of an individual for publication.

This form is especially necessary in the case of talent if the producing company may profit from the eventual sale of the video product. If talent contracts were completed and signed, talent releases were probably included. When talent volunteer their services or perform for only a nominal fee or other gratuity, a signed talent release is recommended. Generally, any talent being featured in video and audio taping should sign a talent release form before the commercial is aired.

*Glossary*

**Talent Name**  This entry should contain the name of the individual talent recorded on video and/or audio tape.

**Commercial Title**  This entry records the title of the commercial.

**Recording location**  The location site or production facility where the video and/or audio recording is made should be entered here.

**Producer**  The name of the supervising producer should be entered here.

**Producing Organization**  The incorporated name of the producing organization should be entered here.

*Note:* The expression "For value received" may imply that some remuneration, even a token remuneration, be required for the form to be legally binding. When there is any doubt about the legal nature of the document, consult a lawyer.

• **Postproduction Cue Sheet Form**

*Production process*  Video commercial postproduction

*Responsibility*  Director

*Purpose*  To prepare an editing cue sheet before postproduction editing. The entries on the postproduction cue sheet are drawn from the videotape log form of acceptable videotaped takes from the source tapes.

*Objective*  The postproduction cue sheet is a stage of postproduction editing preparation in which the director culls from the videotape log form the acceptable videotaped takes and videotape number and lists these sources in chronological script unit order.

*Glossary*

**Producer/Director**  The names of the producer and director are listed here.

**Editor**  Should the director not do the editing, an assigned technical director/editor is listed here.

**Commercial Title**  The title of the commercial is listed here.

**Client**  The name of the client for whom the commercial is being produced is entered here.

**Length**  The length of the commercial in seconds is recorded here.

**Date**  The date of the creation of the cue sheet is recorded here.

**Script Lines**  This column records in chronological order (i.e., in the consecutive order of the master script) all of the script line numbers.

**Log Form**  Every log form was numbered consecutively. Coordinating each log form by number with its respec-

tive script units provides ready access to other information needed during postproduction.

**Tape No.**  The number of the videotape stock on which the script unit was recorded is entered here. This too will facilitate ready access.

**Take No.**  This entry records the acceptable take number on the source videotape previously listed. This information was originally recorded on the videotape log form (and on the continuity notes form).

**Time/Code**  This notation is made from the SMPTE time code striped on the source tapes during or after videotaping for the take listed in the previous column. If time code is not used or is not known, the timer time or videotape recorder counter reading should have been recorded on the videotape log form or the continuity notes form. These readings should have been made from the beginning of the tape as an accurate indication of the position of the take from the front of the videotape.

**Length**  If accurate timing of the length of individual videotaped takes were recorded, that time notation should be entered here. A sum of these entries should give the producer and director a ballpark idea of the overall length of the commercial.

**Comments**  This area of the form allows additional remarks to guide editing.

**Check off Column**  The end column of squares is added as a convenience for checking off the units as they are edited onto the master videotape.

• **Effects/Music Cue Sheet Form**

*Production process*  Video commercial postproduction

*Responsibility*  Audio director

*Purpose*  To list in chronological order all prerecorded audio track(s), sound effects, and music to be mixed with the sound track in postproduction.

*Objective*  This form organizes by script line unit all prerecorded audio track(s), sound effects, and music required for the commercial, listing their sources and duration with the in-cue and out-cue for each.

*Glossary*

**Audio Director**  The name of the audio director responsible for the audio design of sound effects and music is entered here.

**Director/Editor**  The name of the director or videotape editor is listed here.

**Commercial Title**  The title of the commercial is listed here.

**Client**  The name of the client for whom the commercial is being produced is entered here.

**Date**  This notation records the date of the preparation of this cue sheet.

**Page  of**  This entry logs the expected number of pages for this cue sheet and the number of each page of the form.

**Script Line(s)**  This column records the script line number(s) from the master script requiring some postproduction effects or music.

**Effect**   The specific sound effect required is listed here. Any prerecorded audio track(s) should be entered here.

**Music**   The specific music required is listed here.

**Source: Cut/Fade**   This entry notes the recorded source from which an effect or music is to be recorded. The source may be a prerecorded cartridge, a cassette, or a record album. In addition to the source of the effect or music, notation must be made whether the source is to be cut in or faded in and cut out or faded out. Check marks should be made in the appropriate boxes in this column.

**Time**   This column records the duration of the required effect or music, measured from the succeeding in-cue to out-cue.

**In-cue**   This entry notes the dialogue or action cue at which the effect or music is to begin.

**Out-cue**   This entry notes the dialogue or action cue at which the effect or music is to end.

# Glossary

**Academy leader** The academy leader is the first minute of video preceding the content video of a recorded videotape. It consists of 30 seconds of color bars and audio tone, followed by 20 seconds of the slate of the content of the program, then by 10 seconds of black screen, and, finally, by the opening of the videotaped program.

**Action properties** Action properties are those moving objects used by actors and actresses. For example, an automobile or horse and buggy are considered action properties.

**Ambience** Ambience is any background sound (e.g., city traffic or an airplane flyover) in a recording environment.

**A-roll** An A-roll video is the primary videotape recording source. In a studio production videotape environment, the A-roll is the master videotape of the program. B-roll is the videotape source that is inserted into the A-roll videotape.

**Aspect ratio frame** An aspect ratio frame is the television screen proportional rectangle drawing, 3 units high by 4 units wide. Aspect ratio frames are drawings used in the design of storyboards. (See the storyboard form.)

**Audience demographics** Audience demographics is the sum of the individual traits of an audience (e.g., age, sex, education, income, race, and religion).

**Audio perspective** Audio perspective is the perception that longer video shots should have a more distant sound and closer video shots should have a closer sound. Audio perspective attempts to recreate the sound distance perception of real life.

**Audio plot** Preparing an audio plot is a preproduction task requirement of an audio director, which includes the judgment of type of microphone to record required sound, microphone holder for picking up required sound, and physical placement of a microphone for a videotape shoot. (See the audio plot form.)

**Bird's eye view** A bird's eye view is the point of view of a set looking directly down on the set from above, noting the confines of the set and set properties. The view can also contain the cast and the camera(s).

**Bite** A bite is a portion of a video or audio recording actuality.

**Blocking** Blocking is that process by which a director physically moves participants (cast and camera(s)) to differing points within a location or set.

**Blocking plot** Preparing a blocking plot is the preproduction task of a director, which consists of making a bird's eye view drawing of a set or recording environment with major properties indicated. A director indicates with circles where talent will be placed. The circles are combined with arrows to indicate movement of talent. From a blocking plot, a lighting director can create a lighting plot and a camera operator can decide camera set-up and placement. (See the blocking plot form.)

**Boom microphone** A boom microphone is a microphone, usually directional, designed to be mounted and held above the person(s) speaking. A boom microphone must be aimed at the mouth of the speaker and raised and lowered depending on the camera framing of each camera shot. (*See also* Audio perspective.)

**Breakdown** A breakdown is a preproduction analysis of either a script or a storyboard. It is intended to separate scene elements from the script or storyboard and arrange them in proposed videotaping order. A breakdown is a necessary component to the development of a production schedule. (See the script breakdown form.)

**Bridge** A bridge is a video or audio segment that connects, often in summary form, one video or audio subject to another.

**B-roll** A B-roll is a second videotape source needed in production to perform some video effects involving two video sources during videotaping or editing, such as a dissolve or a wipe. The A-roll would be the primary videotape source into which a B-roll source is inserted.

**Cast call** The cast call is that appointed rendezvous time at which the members of the cast—actor(s) and actress(es)—assemble before beginning production tasks.

**Cel animation** Cel animation is the animation technique of painting images on sheets of celluloid. The advantage of cel animation is the see-through quality of celluloid, permitting one under layer painting for backgrounds and other unmoving objects in the animation field of view. Disney animated films are an example of the cel animation technique.

**Character generator** A character generator is a video effects generator that electronically produces text on a video screen. The text that is recorded in the memory of the character generator is usually used for purposes of matting over a colored background or other video image.

**Characterization**  Characterization is the process by which an actor or actress becomes the script character being played. Characterization includes not only costuming, make-up, and dialogue but also the mental and motivational state of the character.

**Claims**  Claims are product or service assertions made to the audience, usually in commercial advertising. Commercials often claim that the product or service advertised produces certain effects. Claims made by commercials should be verified.

**Clearance**  Clearance is the process of securing the rights to use copyrighted material. Clearance is most often secured for the legal use of music.

**Composition of a shot**  The composition of a camera shot indicates the subject and arrangement of a shot as framed in the viewfinder of a camera. It would indicate the person or object to be framed and the degree of the framing (e.g., CU or XLS).

**Consumable props**  Consumable properties are those properties that are consumed as part of using them (e.g., a cigarette is smoked or food is eaten). Consumable props require constant replacing.

**Contingency**  Contingency is that percentage amount added to a subtotal of estimated costs in budget making. A common contingency amount is 15% (i.e., 15% of the subtotal of estimated budget costs is added to the subtotal itself as a hedge against actual costs).

**Continuity**  Continuity is the flow of edited images and the content details of edited images from shot to shot. Continuity observation entails the close scrutiny of talent, properties, and environment during videotaping to ensure accurate flow of edited images in postproduction. (See the continuity notes form.)

**Contrast ratio**  Contrast ratio is the proportion of light to dark areas in electronic video images or across lighted areas on a set.

**Control track**  The control track is a flow of electronic impulses recorded on the edge of the videotape that serve as synchronization units for accurate videotape editing. They serve the same purpose as do the sprocket holes on film.

**Copy**  Copy refers to any scripted text to be recorded on the audio track of a videotape or in the memory of a character generator for matting onto a video image or background.

**Copyright**  Copyright is the legal right of an artist or author to the exclusive control of the artist's or author's original work. Copyrighted material is protected by law and the public use of such material must always be cleared by a producer from the owner of the copyright.

**Cost per drop**  Cost per drop is the manner in which an amount of prerecorded music library tracks is determined. A producer choosing to use an amount of prerecorded library music figures the cost owed to the music library owners by the length of time of the music used (e.g., 10 seconds, 30 seconds, etc.). Rate costs per music drop is part of the contract a production agency has with the owners of the music library. Usually, a payment check need only be postmarked by a mail service before the public use, broadcast, or cablecast of the chosen music drop. This is the convenience of cost per drop music services.

**Crew call**  A crew call is the stated time for the rendezvous of the production members and is usually at the videotaping site.

**Cut-away**  A cut-away is a video of related but extraneous content inserted into the primary video. For example, video images of a hospital operating room (related but extraneous) would serve as a cut-away insert to a video of an interview (primary video material) with a doctor.

**Cut-in**  A cut-in is a video of necessary and motivated video images to be edited into an established or master scene. For example, close-up shots (necessary and motivated) of two people in conversation serve as cut-ins to a long shot (master scene) of the two people walking and talking.

**Cyclorama**  A cyclorama is the ceiling to floor material used as a simple backdrop for some television studio productions. Cycloramas may be made of a black velour to provide a solid black background or a scrim material that may be lighted with any colored light.

**Decibel**  A decibel is a unit of sound that measures the loudness or softness of the sound.

**DVE (Digital video effects)**  Digital video effects are the set of visual image variations created electronically by a control room or editing suite switcher. DVE capability changes analog signals into digital information. Common DVE effects are the image rotation, peel away image, and the mosaic effect.

**Edited master**  An edited master is the final editing of a video piece from source tapes.

**Emotional appeal**  An emotional appeal is a persuasive device used in writing and imaging commercial advertising as a hook to grab the emotions of the audience.

**End board/end slate**  The end board or end slate are terms for the slate when a videotaped take is slated at the end of a take instead of at the front. The end board is used when a take is redone without stopping the camera and recorder. End board use often occurs when an actor simply flubs a line.

**External video signal**  The external video signal is the control room signal that carries the program or line of the video program. In some studio operations, that signal can be routed to the individual cameras and called up by camera operators on their cameras. Having access to the external video signal allows camera operators to match the camera framing over their monitors to the camera framing as seen on the monitor of a camera on line.

**Final audio**  Final audio is used to designate the last sound of copy or music in a video piece. It is a designated end point to measure the length of time of an edited master video piece. (*See also* First audio.)

**Final edit**  The final edit is the completed master tape of a video project. It is usually referred to in contrast to the rough edit, which is a preliminary edit of the videotape.

**First audio** First audio is used to designate the first sound of copy or music in a video piece. It is a designated beginning point to measure the length of time of an edited video piece. (*See also* Final audio.)

**Fishpole** A fishpole is an extendible holder for a directional microphone. A fishpole, usually handheld, is extended into a set during dialogue for videotaping.

**Foldback** Foldback is the process of feeding the audio signal back into the studio for the convenience of studio personnel and on-camera talent.

**Format** A format is the listed order or outline of the content of a video product or program. Format is also used to indicate the genre of a television program (e.g., talk show or newscast).

**Framing** Framing is the composition and degree of image arrangement as seen through the viewfinder of a camera. The framing is usually described in terms of how close or far away the subject is perceived to be from the camera (e.g., a close-up versus a long shot).

**Freelancer** A freelancer is a person who works in the film or video field on a production basis as opposed to being a permanent, full-time employee of a company. Writers, producers, and directors are examples of freelancers.

**Freeze** A freeze in video production is the appearance of holding the video image on the screen still. A freeze is also used for production cast members when they must remain without movement while continuity notes are being made.

**Friction** Friction, usually pan and tilt friction, refers to the amount of drag that the pan or tilt mechanisms produce in performing their functions. Adequate friction gives a camera operator control of panning and tilting motions.

**F-stop** The f-stop units, the calibrated units on the aperture of the lens of a camera, determine the amount of light entering the lens and falling on the pickup tubes.

**Gels** Gels (gelatins) are filters used in the control of light for videotaping. Gels are used on lighting instruments to filter light or change lighting temperature and on windows to change the temperature of light entering a videotaping environment.

**Hand properties** Hand properties are those objects needed (handled) by the talent. Examples of hand properties are a knife or a purse.

**In-cue** An in-cue is the beginning point of copy, music, or video screen at which timing or video or audio inserting is to occur. (*See also* Out-cue.)

**Insurance coverage** Insurance coverage is required for the use of certain people (e.g., underage children or movie stars), facilities (e.g., television studio or remote locations), and certain properties (e.g., horses or automobiles).

**Interviewee** An interviewee is the person being interviewed.

**Interviewer** The interviewer is the person who interviews.

**Lead** A lead indicates the beginning video or audio of a videotaped piece. Leads may be created as a voice-over, a stand-up, or with music.

**Leader** The leader is the beginning portion of audio or videotape that is used to record information about the subsequent video or audio. Most leaders (often called academy leaders) contain the record of the slate and an audio check (e.g., :30 of tone) with a portion of video black before the video content of the recording. Some leaders also contain a portion of the color bars.

**Legal clearance** Legal clearance is the process by which recording and reproduction rights are obtained to use copyrighted materials. Legal clearance must be obtained for copyrighted materials (e.g., music, photographs, or film) and for synchronization (putting pictures to copyrighted music).

**Levels** Levels are those calibrated input units of light, sound, and video that have to be set to record at acceptable degrees of unity and definition for broadcast reproduction.

**Lighting design** The lighting design is the preproduction stage during which the mood, intensity, and degree of light for a videotape production are created. The lighting design is the responsibility of the lighting director. The lighting design can be done after location scouting is complete. (See the lighting plot form.)

**Location** A location is that environment outside a recording studio in which some videotaping is to be done; often referred to as the remote location or, simply, the remote.

**Logging** Logging is the term used to indicate the process of recording continuity details during videotaping.

**Marketing problem** A marketing problem is that challenge in marketing products or services that creates the need for advertising. A marketing problem may be as simple as determining how to get a product known to the public.

**Master script** A master script is the copy of the script that contains the preproduction information for videotaping. The master script is usually the director's copy and contains the final version of the copy, the storyboard, and the blocking of the cast and action props.

**Master tape** A master tape is the videotape containing an edited video project.

**Mike grip** A mike (microphone) grip is the individual responsible for holding the microphone or a microphone holder during the recording of audio in a studio or on location.

**Mixing** Mixing audio tracks in videotape editing is the process of combining two recorded audio tracks into one audio channel. Mixing usually adds music or ambient sound to a track of voice recording.

**Out-cue** An out-cue is the end point of copy, music, or video at which point timing or video or audio inserting is to end. (*See also* In-cue.)

**Pacing** Pacing is the perception of timing of the audio or video piece. Pacing is not necessarily the actual speed of a production or production elements, but it is the coordinated flow and uniformity of sequencing of all production elements (e.g., music beat, copy rhythm, or video cutting).

**Pad** Pad (padding) is a term applied in television production whenever some flexible video, audio, black signal, or time may be needed.

**Pickup pattern** Pickup pattern designates the sound sensitive area around the head of a microphone. It is the area within which sounds will be heard by the microphone.

**Platform boom** A platform boom is a large device for holding a microphone and extending it into a set during videotaping. Most platform booms are on wheels, need to be steered when moved, and require at least two operators—one to handle the microphone and the boom on the platform and the other to move the platform.

**Plot** A plot is a creative or technical design usually used for blocking actors and actresses within sets, designing sound coverage in audio production, designing the lighting pattern on sets, and listing all properties.

**Preproduction script** A preproduction script is a copy of a script that is considered subject to change. A preproduction script should contain sufficient audio and video material on which to judge the substance of the final proposed project.

**Production meeting** A production meeting is a gathering of all production personnel to review details of a production.

**Production statement** A production statement is a simple, one sentence expression of the goal or objective of a videotape production. The production statement is a constant reminder at all stages of production of exactly what is being accomplished and why.

**Production value** A production value is any element or effect that is used when motivated to create an overall impact. Some examples of production values are music, lighting, and special video effects.

**Product shot** A product shot is the photograph or image of the product being advertised shown by itself in the photograph or on the screen.

**Prompter** A prompter refers to either the production crew member responsible for coaching talent during production or the hardware used to project the script copy of a production to the front of the television cameras to assist the talent in reading.

**Property** Property is the term used for any movable article on a set. Properties can include such items as furniture, lamps, telephone, food, or bicycle. (*See also* Set properties, Hand properties, Action properties.)

**Rate card** A rate card is the list of space, hardware, personnel, services, and the cost for production facilities.

**Rational appeal** A rational appeal is a persuasive approach used in writing and imaging advertising in an attempt to grab the attention of and convince the rational nature of the audience.

**Record button** The record button is a circular red insert plug found on the bottom of most videotape cassettes. Removing the record button serves as a safety check against recording over or erasing previously recorded video. The absence of the record button will allow playback, but not recording.

**Roll tape** "Roll tape" is the expression used to signal operation of the videotape recorder at the beginning of a take. A location director uses the expression as a sign of the director's intention to videotape a scene.

**Rough edit** A rough edit is the result of a first post-production editing session where accurate timing and tight edits are not required. A rough edit serves as a preliminary step to the final editing of such tightly edited video genres as commercials. A rough edit provides the opportunity to make final edit decisions on pacing and image transitions.

**Royalty** Royalty is the share of proceeds from the publication of some work of art such as a book, a play, or music. The performance of such works requires financial payment to the owner of the work.

**Script unit** A script unit is that section of a script that designates a videotaping portion and is equivalent to a scene in a larger act. It is any gratuitous unit that a director may define for production purposes.

**Secondary motion** Secondary motion refers to those movements that occur with the movement of the camera. Secondary movements include pan, tilt, dolly, truck, arc, zoom, pedestal, and boom.

**Serendipity syndrome** The serendipity syndrome refers to those good and pleasant effects in television production that were unplanned and unexpected.

**Set properties** Set properties are those objects used to appoint a studio set. For example, furniture, lamps, or curtains are considered set properties. (*See also* Hand properties and Action properties.)

**Shading** Shading in video production is the control of the iris of the lens of a camera. It is the process of setting and/or controlling the iris to permit or exclude light from hitting the camera tubes.

**Shoot** Shoot is a slang term used to describe a videotape production session either on location or in the studio.

**Shooting order** The shooting order is the order in which the script units will be shot and is indicated on the production schedule. Most often, the shooting order is determined by the availability of locations and actors/actresses.

**Shooting units** Shooting units refer to those portions of a television script that are producible in one continuous videotape take. Shooting units in television are similar to short scenes in theater. In commercial production, shooting units are determined by lines of script.

**Shot list** The shot list is a form created during preproduction on which a director indicates types and order of shots to be videotaped. A shot list differentiates between master or establishing shots and cut-ins, provides framing instructions, and indicates the duration of each shot by out-cue. Shot lists are location specific with the shot list for every location beginning with the count of one.

**Slate** The slate is an audio and video recording device that allows the labeling of the leader of each take. A slate usually records the title of the production, the producer and/or director, date, take number, and videotape code. The character generator, a blackboard, or a white showcard can serve as a slate. Slate also indicates the action of recording the slate on the videotape leader.

**SMPTE time code** SMPTE (Society of Motion Picture and Television Engineers) time code is an electronic signal recorded on a secondary audio track of videotape to assist an editor in accurately creating a videotape edit. SMPTE time code records the hours, minutes, seconds, and frame numbers of elapsed time for each video frame.

**Source tape** The source tape is any videotape stock used to record video that will later be edited into a larger videotape project. Source tapes are edited onto a master tape.

**Spike marks** Spike marks are the result of spiking cast or properties during production. A common form of spike marks are colored adhesive dots placed at the blocked positions of actors and actresses. A different color is assigned to each cast member.

**Spiking** Spiking is the mark placed on the set for recording the blocked position of a cast member or a set property.

**Stand-by** Stand-by is a verbal command that indicates that the director is ready to begin videotaping a location unit. Production and crew respond to a stand-by command with silence and readiness to begin.

**Stills** Stills are photographs or 35mm slides of some product or subject. Stills can be created in video with a freeze frame. Stills are used often in commercials to display the image of the product being advertised. They are often taken in the environment in which a commercial is being videotaped.

**Storyboard** A storyboard is a series of aspect ratio frames on which are sketched the proposed composition and framing of each shot to be videotaped. Storyboard frames are numbered consecutively and the audio copy associated with each proposed shot is recorded under the frame. Storyboards are considered essential to some video genres (e.g., commercials) and are encouraged as a quality preproduction stage for all genres.

**Strike** A strike is the final stage of a location shoot when all production equipment is disassembled and packed for removal and the shooting environment is restored to the arrangement and condition found upon the arrival of the production crew.

**Striping** Striping is the process of recording SMPTE time code or control track on videotape as a measure of videotape control during editing.

**Sweetening** Sweetening refers to the process by which audio and video signals are cleaned up and clarified electronically in postproduction. Sweetening audio means to filter out background noises such as hums and buzzes.

**Synchronization rights** Synchronization rights are those legal clearances in which a producer receives the right to use copyrighted music in a videotape production.

**Take** A take is a single videotape unit from the beginning to the end of recording. A take usually begins with a recording of the slate and ends with a director's call to cut. It is not uncommon to record many takes of an individual unit. Many takes may be required for one shot.

**Talent** Talent is the term used to designate any person who appears in front of a camera, including actors/actresses and extras. Even animals are referred to as talent.

**Talent release** A talent release is a signed legal document by which a producer obtains the right to use the image, voice, and talent of a person for publication.

**Target audience** Target audience is the designation of that subset of the public for whom a particular video piece is designed. Knowing a target audience permits a producer and director to make calculated choices of production values to attract and hold the interest of the targeted group.

**Timing** Timing is the process of recording the length of a video piece from first to final audio. First audio and final audio might be music and not a vocal cue. Some video pieces may have a visual cue at the beginning or end of the piece.

**Titling** Titling is the design and production of all of those on-screen visual elements that create the title of a video production.

**Vectorscope** A vectorscope is an oscilloscope used to set and align the color of images as they are recorded by the videotape recorder.

**Videographer** A videographer is a photographer working in video.

**Voice-over** A voice-over is a production technique in which an announcer's voice is heard without the announcer being seen in the video portion.

**Wrap** A wrap is the stage of a production when the director indicates that a good take has been videotaped and signals a move to another shot from the shot list. A wrap is distinguished from a strike.

# Selected Bibliography

## TELEVISION PRODUCTION TEXTS

Armer, A. *Directing Television and Film.* Belmont, CA: Wadsworth Publishing Co., 1986.

Blum, R. *Television Writing From Concept to Contract.* New York, NY: Hastings House, 1980.

Blumenthal, H. J. *Television Producing & Directing.* New York, NY: Harper & Row, 1988.

Carlson, V., and Carlson, S. *Professional Lighting Handbook.* Stoneham, MA: Focal Press, 1985.

Fielding, K. *Introduction to Television Production.* New York, NY: Longman, 1990.

Fuller, B., Kanaba, S., and Kanaba, J. *Single Camera Video Production: Techniques, Equipment, and Resources for Producing Quality Video Programs.* Englewood Cliffs, NJ: Prentice Hall, 1982.

Garvey, D., and Rivers, W. *Broadcast Writing.* New York, NY: Longman, 1982.

Hubatka, M. C., Hull, F., and Sanders, R. W. *Sweetening for Film, and TV.* Blue Ridge Summit, PA: TAB Books, 1985.

Huber, D. M. *Audio Production Techniques for Video.* Indianapolis, IN: Howard Sams & Co., 1987.

Kehoe, V. *Technique of the Professional Make-up Artist.* Stoneham, MA: Focal Press, 1985.

Kennedy, T. *Directing the Video Production.* White Plains, NY: Knowledge Industry Publications, Inc., 1988.

Mathias, H., and Patterson, R. *Achieving Photographic Control over the Video Image.* Belmont, CA: Wadsworth Publishing Co., 1985.

McQuillin, L. *The Video Production Guide.* Santa Fe, NM: Video Info, 1983.

Miller, P. *Script Supervising and Film Continuity,* Second Edition. Stoneham, MA: Focal Press, 1990.

Millerson, G. *Video Production Handbook.* Stoneham, MA: Focal Press, 1987.

Nisbett, A. *The Use of Microphones,* Second Edition. Stoneham, MA: Focal Press, 1983.

Schihl, R. J. *Single Camera Video: From Concept to Edited Master.* Stoneham, MA: Focal Press, 1989.

Souter, G. A. *Lighting Techniques for Video Production: The Art of Casting Shadows.* White Plains, NY: Knowledge Industry Publications, Inc., 1987.

Utz, P. *Today's Video: Equipment, Set Up and Production.* Englewood Cliffs, NJ: Prentice Hall, 1987.

Verna, T., and Bode, W. *Live TV: An Inside Look at Directing and Producing.* Stoneham, MA: Focal Press, 1987.

Weise, M. *Film and Video Budgets.* Stoneham, MA: Focal Press, 1980.

Wiegand, I. *Professional Video Production.* White Plains, NY: Knowledge Industry Publications, Inc., 1985.

Zettl, H. *Television Production Handbook.* Belmont, CA: Wadsworth Publishing Co., 1984.

Zettl, H. *Sight, Sound, Motion: Applied Media Aesthetics.* Belmont, CA: Wadsworth Publishing Co., 1990.

## COMMERCIAL PRODUCTION

Gradus, B. *Directing: The Television Commercial.* New York: Hastings House, 1981.

Wainwright, C. *Television Commercials: How to Create Successful Advertising.* New York: Hastings House, 1970.

White, H. *How to Produce Effective TV Commercials,* Second Edition. Lincolnwood, IL: NTC Business Books, 1986.

# Index